Cyborgenics

the Merging of Man
and Machine

The quickening pace of technological innovation is bringing society in contact with a quality of life historically unprecedented. The transformation of cultures into a globally connected marketplace is taking place at clearly breakneck speeds. Technology can be a double-edged sword; it can work for us or we can allow it to work against us. Time-saving devices can often become devices that help us waste time rather than conserve it.

We have come far, but can civilization further survive the social consequences of embracing a worldview void of moral absolutes . . . a place where right and wrong are not so easily discernable given truth is commonly held as a relative term?

As the grand human experiment continues, many are becoming concerned our dependence on technogadgetry will inevitably lead us to merge with them. Where will we draw the line?

> Cyborgenics helps us decipher the issues of our postmodern age as we contemplate a fast approaching post-human one. Hanson has provided a much anticipated work that satisfies a world hungry and in need for answers to the both new and age old problems facing humanity.

CYBORGENICS

The Merging of Man
and Machine

Ryan Andrew Hanson

Copyright © 2007 by Ryan Andrew Hanson

ISBN 0-7414-3770-8

Unless otherwise indicated, Scripture quotations are from the Holy Bible: King James Version.

Scripture quotations marked (NKJV) are from <u>The Holy Bible</u>, New King James Version © 1984 by Thomas Nelson, Inc.

Scripture quotations marked (NASB) are taken from the <u>New American Standard Bible</u> ®. Copyright The Lockman Foundation 1960, 1962, 1963, 1968, 1971, 1972, 1973, 1975, 1977, 1995. Used by permission. (<u>www.Lockman.org</u>)

Published by:

INFINITY
PUBLISHING.COM
1094 New DeHaven Street, Suite 100
West Conshohocken, PA 19428-2713
Info@buybooksontheweb.com
www.buybooksontheweb.com
Toll-free (877) BUY BOOK
Local Phone (610) 941-9999
Fax (610) 941-9959

Printed in the United States of America

Printed on Recycled Paper

Published January 2008

DEDICATION

To those who never tire or cease in asking life's most pressing and persistent questions...

...who humbly admit they don't know everything—yet nevertheless persevere in the search and discovery of true knowledge and wisdom...

...whose ends of knowledge is to benefit the whole of humanity—not, rather, for mere personal satisfaction, political gain, and/or self-glorification...

...who are grateful for this life—for the chance to live, to love, to be loved, and to ultimately challenge the status quo—contributing to the development of a world where our children may reap the blessed fruits of our purposeful labors...

...and to the Source of all things seen and yet to be seen, making every opportunity possible along the way.

This book is for all of us.

"May we seek to know more about each other, ourselves, and our ultimate purpose in the search for and communion with truth, however apparent or elusive."

~ Ryan Andrew Hanson
2006

ACKNOWLEDGMENTS

For reasons known to my friends and family, more obstacles than expected had to be overcome in the research, writing, and publication of the work you now possess. Whether received as a gift or purchased, it is undervalued compared to the efforts of the following contributors, from whose fruit you will ultimately benefit. All of them deserve recognition and honor for their parts in making this project a success.

Frederick W. Hanson, Jr., M.D.:	words fail to describe the love you show
Mike & Wendy Barnard:	for support without need of asking
Terry & Allison Hudson:	for interest and concern in earnest
Warren & Connie Taylor:	for going to bat and making contact

Tom & Marijen Hudson: for keeping me informed and in prayer

Barbara Huval, Ph.D. (Rice):   what joy for language, learning, and life you bring!

Ravi Zacharias: for more inspiration than you know—to reach this generation

Tom Harvey: my missionary friend and brother

Laurah Murphy: my favorite Canadian

Kim Binkley Smith: you were there when others were not

Michael "Cowboy" Nunn: ...to know Him and the power...

Jonathon Swanson, Ph.D.:   a true Renaissance man of our age

David L. Botsford: always giving me your best

Chris Zavatzky: you came through! Keep in touch...

Hayley Hanson: hoping you find the Answer to your life's questions

Kylie Juarez: always a pleasure...

Michael McCasland: our families are
                            strengthened through
                            our faith and hope

Sue Wright: recognition is where it
                            begins

Sally Byrd: the reward is in the
                            result

Mary Jo Rafferty: there are no small
                            parts

Lloyd Warren: ...you too, brother!

The Infinity Staff: a wonderful family

NOTE TO THE READER

As we each live out our days, time is revealed to be increasingly *of the essence*. It is our most valuable commodity. Rather than cover this work's subject matter extensively, I gathered the most progressive, interrelated materials in their respective fields and strove to hit the high points, while concurrently placing emphasis on keeping all issues discussed pertinent and in proper context.

Great effort has been made to keep every element timeless and relevant in nature. Still, the rapid pace and rise of new technologies will render some areas obsolete in the years that pass. Internet sources as well as periodical references are therefore included and recommended (to remain current on the latest technologies). Thankfully, the moral and ethical standards will forever hold constant and enduring.

More and more often, I see writers veering off on tangents to fill unnecessary pages with irrelevant peripheral matters. As an avid reader myself, I don't take kindly to authors involved in such shenanigans, wasting the precious moments we can never again relive—claiming to provide us with answers they never had. You'll find in this work:

1. Clear efforts to zone in on and make plain neglected or forgotten common knowledge that is essential for us to gain or regain to further progress as a people.

2. Key high points from relevant source materials.

3. No self-promotion of or for any other personal works.

Hard to imagine how some of the leading scientific minds and authors can miss one of the more important goals of modern technology—

EFFICIENCY!

~ R. H.

CONTENTS

Appendixes

PREFACE

New and emerging technologies continue to spawn optimism about eventually realizing an increasingly utopian society. The trend towards globalization, however, has clearly shown us that for such an ideal civilization to exist, we must first overcome current sociocultural barriers in order to progress further.

Understanding why traditional ethnocentric ideologies are no longer conducive in a multimedia connected world is crucial if we are to transcend divisive issues, such as race. Propagating such irrelevancies in a maturing and beneficial world, which is headed towards a more homogeneous global culture, only serves to further hinder our progress in attaining that goal.

If the very laborsaving technologies we embrace are used to encourage and empower individuals to become godlike egocentric megalomaniacs and/or demagogues, a unified culture will be extremely difficult to realize and may easily serve to divide us further, defeating technological purpose altogether. There can be no unity with *this* type of mass diversity.

Dualistic or partisan competitions (in sports, religion, politics, etc.) become misguided in such cases. The agendas become power-grabbing and self-seeking in motivation. Unity toward universal benevolence is non-existent where everyone sets their own standards. Seemingly sincere perceptions of reality can often be sincerely *wrong*, and must be confronted if world peace and progress toward a more perfect global community is earnestly desired. Avoiding problems only complicates them further when the ultimate goal is to solve them—they must be confronted and openly discussed to reach full consensus; surely difficult and time consuming, but not impossible. This work aids to simplify what is deemed necessary yet seen as too difficult to tackle and overcome.

Relevant dialogue that finally disposes of the "politically correct" mantra/monster (which appeals to and sets up pure emotionalism as a standard for morality) and proceeds to get at the root of individual beliefs will unearth each of our most cherished desires, so we can better position ourselves in society where our unique talents and skills are used most effectively. We are all unique and gifted in areas we love and enjoy; those skills serve the whole when implemented without exploiting the individual, and emphasize his or her essential worth. Quite often, however, labor is used to achieve or prop up some demagogue's twisted worldview. We have plenty of those already...what we need is a firm foundation to build upon, to escape our flawed natures and tendencies to skip proper groundwork and build upon shifting sands.

There comes a time when it is neither safe nor wise to assume that everyone we encounter knows what is right and good, when the standards are so much in question and so far from common knowledge. Our dialogues must begin to include a way to discover, inspect, and—if need be—reassert the foundations of well-established standards that nurture our conscious humanity for the common good.

There is no other agenda in this work but to assist the reader in how best to begin helping others know and apply the true standard for morality and belief, so as to avoid falling victim to the present culture that promotes the enigmatic "relative truth" doctrine (the quintessential oxymoron). Demagoguery thrives in such subjective environments, at the expense of ignorant masses oblivious to its underlying, self-serving motives.

Misologists may despise the following chapters, and miss entirely the primary goal—increasing overall awareness of the sociocultural forces at play, thereby increasing the ability to escape self-delusional manufactured realities, and unleashing the creative potential to both discover and attain your truly-intended abundant life and purpose, which surpass any present utopian ideologies. Joyful living *is* within your grasp!

A lifetime student of ceaseless learning, I cannot help but concede the elusive character of language in efforts to adequately present thoughts and ideas during these transient times of rapidly evolving cultures.

In effect—to the degree that the body of this work breaks from traditional literary form or presentation—the design of the text is quite purposeful.

It would be crass contemplation to conclude that every man has a full grasp on where our generation fits into the historical context of civilization.

To reach the widest of audiences—to speak to every man, woman, and child—bringing depth of understanding to trends affecting how we live and view life in general, a "break" in traditional form serves quite productive for obvious reasons.

For this reason, connoisseurs of linguistics and other purveyors of proper syntax are encouraged to condescend and accept conspicuous literary *faux pas* as "necessary evils"—a means to an end toward the common good.

Redundancies abound to both emphasize and enhance meanings or ideas which may be new and therefore vaguely familiar, for instance.

As our standard of living is always in flux—evolving and being redefined constantly—it's not "an inconvenient truth" to connect our growing inability to tolerate extreme summer temperatures with our increasing dependence (and desire) for air-conditioned environments year-round. It's examples like this that should serve as reminders, and put us all on notice that misperceptions of reality are profuse—not to mention hawked—at every corner, under the guises of "community service," "environmental protection," and the like.

Discern and embrace the relevance of content, rather than act as supercritical reviewer of form or style. Add to the discourse—this is not a zero sum game...We are all, proverbially, in the same boat.

Escape the common tendency to destroy opposing views; rather, seek to understand them and even accept their precepts when honorable to do so. Diehard adherents of destructive criticism are encouraged to ponder Malcolm Muggeridge's epigram of whether "we have educated ourselves [as a whole] into imbecility," and also to take a gander at David Foster Wallace's *Harper's Magazine* review of *A Dictionary of Modern American Usage* to satisfy any appetite of the hypercritic or cynic. (Closest thing to a plug you'll find from me)

Thinking the seemingly difficult issues through will both serve to both clarify as well as *simplify* them.

Let's now enter on common ground...

Knowing that—

> *"Not only do life's answers at times end up right before our eyes...they never cease and will always be with us wherever we go."*
>
> ~ R. H.

"In times of change, learners inherit the earth"

~ Erich Fromm

INTRODUCTION

This work has been one of discovery and deep concern: concern for those so wrapped up in the daily grind—making ends meet—that the ever-changing landscape of global culture leaves them in the dust. One of discovery in that a to-the-point writing focused on the main issues of our day can serve as an effective rallying point, one that informs and equips all who share similar concerns, to both bring to light neglected issues that matter and take the opportunity to educate those less informed of them.

To that point, it has been a labor of love, not one of tedium or out of any financial necessity, but one to spark true interest in those who desire to make a difference and influence a generation facing unprecedented challenges, in regard to the increasing part technology is playing in all of our lives.

Above all...*enjoy* it! May it be to increase your awareness and knowledge base of the sociocultural climate, or to simply take the opportunity to discuss meaningful issues with loved ones and friends while stuck in traffic, waiting in line, or at the dinner table. Whatever the setting, after reading even a few chapters you will find yourself in a better position to guide the conversation along a more constructive course, avoiding the common paths toward superficial and/or argumentative ones.

The following chapters are meant to encourage and inform, not cause anxiety and dismay. Be aware of your audience, be it one or many. In an age such as this, you will discover many who often move so far off topic that it will be obvious your words on the issue at hand are taken out of complete context. Be patient. Begin again, re-explaining in full context whenever necessary. Do further research if you see the need. This work is by no means exhaustive on the issues discussed, but merely touches on the most important aspects and implications of them.

Quite inevitably, when those with whom I work and interact find that I am also an author, the question emerges:

"What do you write about?"

My response is mostly similar in content...I write on the most crucial issues of our time, to inform and, hopefully, educate as many as possible about our exponentially-increasing, ever-changing global culture.

"Well, what issues are those?"

Interest is sparked at this point already, so I have a chance to list them, and do an on-the-spot survey at the same time to pinpoint specific interests. To avoid the common "multiple choice" method of picking *A* or *D* (first or last)...I put the issue of *first principles* somewhere in the middle. This issue is almost always (9 out of 10 times) chosen as one of most interest—no matter what order in the list I choose to place it.

What is most compelling is that they are choosing the most *foundational* of issues, which is essentially critical to best understand every other one. After all, it is where our first principles lie that determine our worldview (how we see the world), which in turn establishes our beliefs, and subsequently influences *every single* choice or decision we make from moment to moment.

It is, therefore, *crucial* that we take the time to discuss this issue in the first chapter—however controversial.

Prologue

In order to better reveal the remarkably profound implications behind a man's transformation into a cybernetic organism (cyborg), the lead up to those focal chapters deal with neglected foundational principles, meant to serve as much-needed retrospection.

It has only been in recent decades—and mainly due to our self-manufactured environments—that we have begun to show blatant disregard for the imperative nature and function that philosophical disciplines (such as semantics and ontology) play in overall cultural formation. Doing so, and remaining in our self-made comfort zones, it becomes second nature to label them irrelevant...That is, until the social fabrics of civilization unravel into an unmanageable mass of total anarchy. Of course, by then such studies would be an act of futility.

There seems to be no better time than the present for each of us to help secure a prosperous future, and prevent such highly likely cultural degeneration from progressing any further.

In a period when conspicuously immoral means get a pass to achieve deeply desired ends, a reexamination of our core values should come as a welcome act to at least a concerned remnant of society.

Part one:

FOUNDATIONS

To Freedom

"The ultimate determinant in the struggle now going on for the world will not be bombs and rockets but a test of wills and ideas—a trial of spiritual resolve: the values we hold, the beliefs we cherish and the ideals to which we are dedicated . . . So let us ask ourselves: What kind of people do we think we are? And let us answer: free people, worthy of freedom and determined not only to remain so but to help others gain their freedom as well."

~Ronald Reagan

"Life is a test and this world a place of trial. Always the problems or it may be the same problem will be presented to every generation in different forms."

~ Winston Churchill

CHAPTER ONE

First Principles/First Cause

Controversial it is. Big bang? Evolution? Intelligent design? Prime Mover? Where did it all begin, for what reasons, and how can I know with any certainty?

The debate has never been settled—essentially because all sides have agreed that any of the theories can neither be proven nor disproven.

A convenient conundrum. So why take sides? Simply because our foundational beliefs do *determine* the paths we decide to walk on through life. The best example I have found to illustrate this point is to present the atheist ("there is no God") and theist ("there is a God"). Every choice an atheist would make (if true to belief) *would not* be affected by any accountability he or she might feel to a supreme and intelligent Creator, who would render perfect judgment on creation. A theist or "believer" would. One would have a definite immutable standard for morality or ethics, while the other would not.

As I found the issue of first principles of most

interest in informal surveys, I wondered if it was the subject of God that brought the attraction. Surely not. The word itself usually brought evasive rhetoric in any other context, and the subject was changed immediately. What I found was that it was the *mere possibility* of the existence of a perfect Creator God—which avoided the common and relevant "proven/disproven" argument already discussed, and the common aversion of many to accept any absolutes.

"To suppose that the eye with all its inimitable contrivances for adjusting the focus to different distances, for admitting different amounts of light, and for the correction of spherical and chromatic aberrations, could have been formed by natural selection, seems, I freely confess, absurd in the highest degree."

~ Charles Darwin

The mere fact that all sides of the debate agree on the point that the existence of such a God *cannot* be disproven, leaves room for a reasonable and logical discourse on the irrefutable possibility of existence. On a side note, I have never met a true atheist who could back up their belief...they simply cannot *prove* their position. The ones with whom I *have* had logical discourse ended up conceding to being agnostic (believing they cannot *know* if there is a God, one way or another).

Blaise Pascal, a 17th century French physicist, mathematician, and philosopher (after his conversion to Christianity), gives the following brilliant and insightful illustration in his famous 'Wager':

"Is Christianity true? Why take the chance?"

If Christianity is false, you have gained *little* except a few sensual pleasures as a nonbeliever; as a believer, you have only lost a few sensual pleasures and gained the hope and stability of the Christian lifestyle.

If Christianity is *true*, the nonbeliever would *lose* everything, while the believer would *gain* everything.

What is the wisest decision (wager)? Bet on God.

I have found no other more convincing argument for living in accord with the possibility of such exclusive worldview being true. In lasting and eternal respects—it's a win/win proposition.

A Perfect* Being

How often have you contemplated what the existence of an absolutely perfect Creator Being, or "God," would imply? Such is the purpose of theology—more specifically, the doctrine of God within the field. Number one in importance is to come to grips with the fact that accepting this existence is not dependent on any *belief.* You are reading this book...that is a fact, an

---

* I use <u>perfect</u> only for emphasis – God is by <u>essence</u> perfect.

established truth, a genuine reality. Now, imagine any person you know walking up to you right now and telling you (while you are reading, no less!) that you are *not* reading this book. How absurd is his or her assertion? Would you even spend time trying to prove such absurdity false?

Yet this same method of attempting to establish what is true or factual is not only used but accepted as legitimate in practice!

Anytime someone begins a statement with "I believe," they are often proclaiming that their belief in whatever subject is being discussed establishes the truth of the matter.

National Public Radio produced an entire series of essays, written by both well-known personalities and common citizens and read on air, devoted to the activity. The name of the series? *This I Believe.* (Just a hint at just how ingrained in culture this ideology can become...being fully convinced such activity is prototypical and is to be universally encouraged!)

Is God dependent on or defined by what creation *believes* of Him? Just who is doing the *creating* here?

One could say..."I believe God is holy" and be correct because He is, but not because one merely believed it to be so. God is *holy* because He *possesses* that quality (and many others).

Truth = Genuine Reality, Factual Existence.

Is God..."truth"? We have already established that His existence is quite possible and cannot be disproven. If the existence of God is a fact or genuine reality, then, yes—God would be the ultimate expression of truth. Therefore, consider the following as accurate derivatives of the same first principle discussed with God being the First Cause:

- Beliefs do not establish truth

- Truth establishes beliefs

A perfect Creator Being we call *God* must, by necessity of the name, possess an all-knowing (omniscient), all-powerful (omnipotent), and ever-present (omnipresent) capacity to claim such Deity. To know all, or possess omniscience, *alone* should be sufficient to break one down into a state of utter humility, for then one would be acknowledging that this Being *knows* every thought that has ever entered our minds—and, much more, knows the thoughts that *have yet* to enter them! This is nearly incomprehensible for us to imagine—but, not quite...What about the *actions* we choose to take as a result of those thoughts? Oh, yes...God would know them, too.

Personal Responsibility

Something of a rarity in our time...We all, at one point or another, tried to blame our society, our parents, our enemies, our friends, our genes—whatever is necessary—to throw off any guilt we might incur by accepting that, yes, we—individually—are responsible for our actions. *We* choose our beliefs, *we* choose to act according to them (or not), and *we* must accept the consequences for doing so...whatever the outcome turns out to be, whether grueling or even quite satisfying. Blaming others or circumstances is like throwing a boomerang—you'd better watch it, because the blame comes right back to *you*. Don't get hit in the face claiming ignorance. Now, at least, you'll know better (thank me later).

Another reassuring aspect of a "perfect" God is that possessing the previously discussed qualities also implies that He is *immutable*, or unchanging. Wow! Someone we can count on not changing at any whim! How refreshing (no sarcasm intended, but present nonetheless)!

A Sure Foundation

Being assured of the fact that God cannot change gives us a *standard*...The standard for truth, of course, and subsequently a rule of measure for what is morally and ethically pure. Morals? Ethics? Right and Wrong? Good and Evil? You got it! The very topics also avoided like the plague...and why not?

With moral relativism—the religion of our time—who's to say what's right and what's wrong? Well, we've already dismissed this fantastical worldview. We know better, right?

Okay...for those still holding out, here's an ultimately extreme scenario to make sure we can all discern between good and evil:

You're at home watching your seven year old open her birthday presents; while you and your spouse savor the moment, two burglars suddenly overpower you, knocking both you and your spouse unconscious...you both awaken bound and gagged, watching helplessly while the two kidnap your daughter, only to hold her hostage until you pay their ransom..

Good or Evil? You make the call.

As I'm sure you'll agree, the idea of moral relativism/relative truth is inane. The terms are completely oxymoronic. Otherwise, we would lack any sure foundation on which to base any decision, and each of us could determine what is true, giving any reason to justify our actions. This would leave no choice, for anyone assigned to sit in judgment, but to accept any reason given for their questioned actions as valid and morally acceptable. Sound familiar? Anarchy is not very far behind society embracing a rule of law promoting a doctrine as this.

Truth is *foundational*—an unchanging standard genuinely exclusive by nature, and, therefore, the basis for any rule of law seeking to render equal and fair justice.

So how is this all relevant or applicable to my life right now, you ask?

Completely and absolutely relevant! Each and *every* decision or choice we make is *determined* by what we believe is truth, and if that truth is inseparable from the essence of God as Creator, then we are accountable for *all* of our actions. Given this, would He fail to render absolute justice to a creation that has freely chosen to function outside of their specific purpose by design?

Can we truly know our intended purpose?

Absolutely. Look at reverse engineering. Computers are *designed* for a *purpose*. Their purpose is to operate in the most efficient way by which the designer *designed* the machine. This is true from the CPU to network configuration and to the World Wide Web itself (I discuss this universal fractal-type pattern of design in a later chapter). Cell biologists do much the same thing—observing cell function to determine the nature of their efficiency, in hopes of either improving upon it or reproducing the design. As a result, much has been discovered in recent years to give insight into cell design, and how it relates to our own purposes as individuals who carry out tasks as we relate to our environment—just as cells do. Science, more and more, is giving evidence to an Intelligent Designer/Prime Mover we know and/or label as *God*.

Summary

Already we have touched on various fields such as theology (doctrine of God), philosophy (Socratic method: *if* this, *then* this), technology (computer/internet infrastructure), and biology (cell function). Again, not exhaustive, but rather introductory, in order to show the definite "consequences" of them, and display their interconnectedness.

This is a main purpose for this writing—to discuss these issues, gain a more lucid vision of how they interrelate and contribute to the postmodern technological convergence, which will result in an actual post-human reality—the emergence of the cyborg...the attempt to challenge or extend our own mortality.

We will get to the birth of this age, consummated with the first cybernetic human toward the last chapters...For now, let's seek to better understand the sociocultural dynamics in place today, to understand *cyborgenics* and all it implies contextually.

"Science can not solve the ultimate mystery of nature because in the last analysis we are part of the mystery we are trying to solve."

~ Max Planck

"Humans evolved their cognitive abilities not due to a few accidental mutations, but rather from an enormous number of mutations acquired through exceptionally intense selection favoring more complex cognitive abilities."

~ Bruce Cahn

*"Find out just what any people will quietly submit to and
you have found out the exact measure of justice and
wrong which will be imposed upon them."*

~ Frederick Douglass

Chapter Two

Popular Culture: shaping and reflecting who we are

Does popular culture shape and mold us as individuals? *Is* it a reflection of who we are? Or do we simply find the allusion descriptive and to the point; accepting the degree to which whatever is deemed *popular* by the masses is acceptable behavior for society?

When faced with the issue of what role popular culture ultimately plays in society, we will inevitably decide which side we fall on for ourselves. Considering our current transition into a more homogeneous and accepting society in respects to all prevailing cultures (when the lines between them become less clearly defined), where we stand on the issue becomes of crucial importance for us to adapt to our ever-changing environment without compromising moral or ethical integrity. Information technology brings us closer and closer to cultures and peoples with which we once only infrequently came into contact, in leisure as well as business. Now, the opposite ends of the earth are within an arms reach and without the slightest delay—thanks to super high-speed internet connectivity. "Real-Time," as this instantaneous dynamic is called, whereas before we settled for the time lag of long distance phone calls—which resulted in cutting off one another in mid-sentence at the slightest of pauses! Such days are over. Yes, *technologically speaking*, popular culture is shaping and reflecting who we are. What once restricted us has now set us free, to get on with reaching the goals we dreamed of attaining. The communication barriers have been broken—bringing the ability to network with whomever, wherever, whenever in order to finally let the world know who we are and what we stand for. So what is everyone waiting for? You would think there would be at least a few more geniuses out there with some equally genius ideas on what would be the best direction for our globally connected civilization. Every ship needs a captain, right? Well, not *necessarily*. This spaceship Earth seems to be moving along in this solar system quite adequately, unless someone can prove otherwise…

Seriously, though, we speak of the sociocultural direction of this bright blue planet with a population of *Homo sapiens* now exceeding six billion. There are two sides of this popular culture issue. You have seen the cutting edge of technological advancement in action for yourself in just one decade. The other side is not quite as sharp of an edge; in fact, this edge of the sword needs to be honed if it is to match its razor-sharp counterpart.

What modern society is experiencing is cultural shock from the sudden impact of such unprecedented scientific advancements in information and telecommunication technologies. Quite suddenly, much of the industrialized world (some even call us *civilized*) has access to any and every kind of information known to mankind—from the beginning of recorded history to the present. No longer are we misled on any topic of discussion, having all the facts at our fingertips *within seconds* now. No longer do we have to hold our tongues until we research facts at conventional libraries to refute or confirm a speaker's assertions. Instantaneous confirmation of fact or fiction! How can one go wrong? It *would* seem that such availability of information, being a welcome boon, should act as a catalyst ushering in a new age of enlightenment. In reality, however, without hurdling a few obstacles in our path, society could very well crumble into a chaotic mass of well-informed individuals without the ability to agree on anything whatsoever.

> *...modern society is experiencing a culture shock from unprecedented advancements in technology.*

How so? *Prevailing culture is prone to mistake information for actual firsthand knowledge.* None of us possesses knowledge until we have applied the essential information in the specified area of focus. Knowledge is related information applied. This is necessary for us to understand. If you listen around, you may notice how many conversations, in the media as well as in social circles, are reduced to mere echoes of popular ideological clichés from so-called authorities, which claim to be bearers of truth. Many people believe such assertions because they sound logical, appear rational, and are widely spoken and accepted. The problem is that none of these determine or solely substantiate what is "truth." What is more, knowledge is not even an end in itself—it can result in either foolishness or wisdom, depending on its application. Knowledge and/or truth claimers must inevitably reveal their sources, the standard from which all such claims must be measured. Without a standard or source of asserted objective truth, all claims to such are groundless, having no foundation.

Contrary to *popular belief*, believing something does not make it necessarily true. Something is true because it possesses that quality of existence or being. For example, if I show John a zebra and he calls it a horse—him *believing* it is a horse does not change the fact that it is a zebra (regardless of abnormal vision). Such is a picture of where we stand in social interactions. John would, in fact, be wrong to call a zebra a horse. More importantly, to bring such genuine truth to his attention does *not* ridicule him, *nor* does it make *me* unjust in any way for doing so. No matter how long or regardless of how many people choose to believe something—they do not establish what is genuinely true, either. In ages past, all but a few actually *believed* the world was flat! Did that *make* it flat? So strong was this overriding belief that some of its adherents were more than willing to put to death anyone not sharing the same worldview.

These emotional hang-ups, in connection to beliefs, are not therefore new. No one particularly enjoys being found to be in error, much less having someone point it out to us or others. But which is worse—allowing one to continue in error, risking life and limb in some cases, *or* caring enough to point out irrefutably erroneous beliefs, regardless of the temporary emotional scars such acts of mercy may cause? The latter is no less than throwing a life preserver to a person drowning due to lack of swimming ability! Who would do anything on the contrary? Yet, that is what we do when we refuse to point out errant truth claims in an attempt to preserve another person's emotional stability.

Many choose not to *offend* others with genuine truth bringing great long-term consequences. Such a lack of concern for other s ultimate welfare is where we stand so divided as human society. Even in the most tragic of earthshaking events, people's motives are questioned as mere opportunities to gain public fame and achieve celebrity status! Come, one and all—who will be our next hero or heroine!?! It's becoming rather surreal to witness firsthand… Certainly, a brave new world has emerged to some degree; whether it turns out to be for the

> *Prevailing culture often mistakes information for actual knowledge.*

better or worse will, to a large proportion, depends on us. There is no question that all of us individually have the ability to make a difference. The question is—are you concerned enough about our direction to influence culture for the benefit of all mankind? If so, the time to do just that is now…the progress of society is dependent on its ability to pass on lessons learned by those who came before us.

Technology has freed us to finally dedicate more of our efforts to reevaluating where we are going in this ever-increasingly connected global community. A major obstacle is that today's popular culture seems preoccupied at the moment by satisfying sensual urges through various entertainment media—as if "feeling good" is the ultimate *end* to all of our means. Hopefully, most of you have taken notice of the above trends and are fed up with this "pleasure principle"—a never-ending search for immediate gratification, that only leaves us disappointed and wanting more. If so, you may begin to ask yourself and find the answers to the following questions:

- What am I doing to make this world a better place for our children to live?

- What do I really believe about this world—and where we are going as a people?

- Why do I believe the way I do?

- Can I logically and reasonably give an answer to *others* about such beliefs?

- Would I admit to myself if I was in error of those held beliefs?

- Just what *is* the standard of truth—which founds and ensures right beliefs?

If we encourage ourselves and others to ask questions similar to these—we may find the answers taking us where we should be headed corporately as a people. In the meantime, we will continue to be a *technologically connected and socially disconnected* people, expecting a leader to tell us exactly what to do, yet becoming resentful when he or she actually does so!

Much of our social stumbling is rooted in an innate desire to be accepted into certain peoples' associations or groups. We most certainly will not be accepted by everyone—but if we are honest with ourselves and others, those with the

> *...all of us individually have the ability to make a difference.*

same virtuous character *will!* The potential for mankind to achieve even greater unprecedented heights of human civilization could be realized if we would cast off quirks that only add to the current sociocultural snafus. Your contributions may not be considered "popular" at present, but they may prove vital to tomorrow's culture and beyond. Never underestimate the value of your being. Our *value* is inherently beyond measure when in accordance with our specific human design—and if *designed*, then for a *purpose*.

"A civilization which develops only on its
material side, and not in corresponding
measures on its mental and spiritual side,
is like a vessel with a defective steering gear..."

~ Albert Schweitzer

"So now, from this mad passion—which made me take art for an idol and a king...I have learnt the burden of error that it bore me—and what misfortune springs from man's desire...The world's frivolities have robbed me of the time that I was given for reflecting upon God."

~ Michelangelo

CHAPTER THREE

Is It Entertainment Or Propaganda?

There can be little, if any, doubt that the mass media is a major player of social and cultural influence in today's world. Very few homes, schools, or other institutions within industrialized nations can be found lacking a television, radio or computers linked up to the internet. Millions upon millions of megabytes of information at our fingertips! It is no wonder a substantial number of avid watchers, listeners, and users exhibit short-term attention spans, memory lapses, and other key diagnosing symptoms labeled as Attention Deficit Disorder (ADD). How can the average friend, colleague, or passing acquaintance compete with the professional attention-grabbers employed by ace advertisers, whose primary goal is to drive product sales through the roof? Is it any *wonder* that our own communication skills are sinking to new lows? Ponder the fact that our leading politicians are becoming increasingly dependent upon teleprompters in order to get their messages across with spectacular oratories written by paid speechwriters; yes, especially in the White House. Notice the tremendous amount of energy and funding implemented to produce a mere *message*. Make no mistake—there can be unbridled persuasive power in well-spoken and strategically positioned transmissions of information.

It comes just about natural for most of us to make a goal of purchasing the latest model computer or flat panel television (as if it were *morally imperative* to own them!). For the most part, these gadgets are the primary programming devices of the media empire (along with the motion picture industry, of course) and are what a good number of individuals identify with as the main sources of so-called "entertainment." Buried

> *Who determines what is entertainment...?*

beneath the surface of mainstream America lies one of the deepest roots of social and moral decay, which is going almost unnoticed by those caught up in the latest shapes and forms of *accepted* entertainment in this postmodern world. *Who* determines what is entertainment, news, or true educational programming?

Can the masses even tell the *difference* between subjects being broadcast simultaneously, whose programs are many times viewed mid-program without any introduction and intermittently interrupted by commercial advertising on most networks? One would hope so, but the direction that programming has taken, with increases of violence, profanity, and sexually explicit subject matter—with little public outcry—speaks otherwise. What this passive acceptance of immoral programming says about our culture is that a good portion of our population has forgotten the once supremely valued moral standards set down by our founding predecessors. Shall we now push aside our traditional family values, which promote close-knit family relationships built on trust, and accept this new hedonistic lifestyle that thrives on individualism and materialistic gain? This is no doubt dangerously unstable ground that puts little value on the importance and necessity of the family unit, which is generally accepted as the foundation of any civilized society. If family is observed as a microcosmic model, then civilization is the macrocosm.

Although we have a *choice* in what channel, program, or movie we watch, we can be sure we have little *voice* in the message being sent out by the major broadcasters run by huge conglomerate corporations, entangled with the most bizarre forms of interdepartmental politics imaginable. What we are experiencing is nothing less than a power struggle for *political correctness.*

Can anyone describe this all-inclusive philosophical standard known as the state of being politically correct? It strives to lead us into a utopia where no one offends anyone and truth is subjective. The only problem is that this can never be, since truth ceases to be truth if it is not objective and absolute! *Relative truth* would suggest *no* standard whatsoever, which would pose somewhat of a problem in determining *right* and *wrong* behaviors! Since this doctrine is so prevalent in our present culture and promoted by these mega-media outlets, it would stand to reason that some of us try to return to traditional values in outrage. If our crumbling social-moral fabric is not obvious to all in this age, where the very institution of marriage is being questioned and redefined, then just watch and see what comes next.

The shock value is lessened as we watch the latest *taboo behaviors.* The incremental strategy of the "marketeers" is to increase ratings as discussed. After a while, we may be so desensitized that absolutely *nothing* will shock us. *"They wouldn't show it on television or in the movies if it was not okay for us to see, right?"* And here we are—in the very crux of the real problem of *accepting purely fictional storylines* (however idealized), produced and directed for profit, *as sociocultural norms.* See it for what it is—*propaganda programming*, in effect influencing the masses with an agenda

> *...the very institution of marriage is being questioned and redefined...*

(economic and/or political in nature), with the desired goal of us adopting a set of ideals and to push the envelope as to what we will accept without feeling exploited. Sorry to inform you...that ship sailed some time ago. Those who do not accept this type of "passive aggressive programming" (as the masses surely do) as something *purely* for entertainment value, and choose instead to speak out against this out-of-control downward spiral towards an apathetic, immoral culture, are *immediately* categorized and stereotyped as *conspiracy theorists* or some other oddball extremist who should be silenced.

So, it is ultimately up to *each one of us* to determine what the main message a program or movie is conveying, and what worldview it is attempting to promote. What if the worldview is contrary to the promotion of our core traditional values, which are so essential to sustaining a growing, thriving society? Our core values should primarily be those that strengthen family relations, encourage the building of tightly knit communities (that support one another toward a common goal of the same), and enable people to secure liberty for all present and future generations. Such attainable family and community relations are the model of American society—which was once unquestionable in identity. When the mainstream media crossed the line from *entertainment to propaganda*, which only serves to corrupt our character, should that not be a clear signal to do something about it...else we fall victim to the set agendas of these media institutions? Are we not morally and socially obligated to take a stand against such incrementally corrupting mainstream propaganda, or do we simply continue to be barraged left and right by this type of incessant media blitz? If the latter, we will passively continue to condone (through our silence) a widespread, common, and spreading culture of immorality. Continue to watch without acting (to your peril), as those around you slowly conform to the standards of morality set by the programming they *choose* to submit to. "It's just a movie," is not a viable pretense or excuse anymore as a response to this progressively sophisticated medium of information, which is being used to forward tremendously destructive worldviews that further alienate us from one another—step by *incremental* step.

> *...mainstream media has crossed the line from entertainment into propaganda...*

It is no longer *just* a movie. The motion picture industry represents the pinnacle of entertainment media. Millions of people worldwide are drawn into local movie theaters by the most extraordinary attention-getting methods and tactics ever designed, only to be presented with the most bizarre, violent, morose, degrading audio and visual messages—representing equally deviant worldviews. People go to theaters supposedly to be entertained, but are instead subjected to both blatant and subliminal messaging. The advertising hype is meant to shock and *draw*, pushing adrenaline levels to record highs, no different than the effects of artificial or natural narcotic stimulants...and at what cost? Societal collapse. *Must we again learn* the lessons of the fall of mighty Rome...to be doomed to a similar fate of internal followed by external destruction? History itself beckons us to learn from her wisdom, or grieves for us to repeat needlessly the same painful lessons over and over.

The moral obligation of cinematic producers and directors, in particular, has never been greater. Digital technologies are able to transport the viewer into an almost virtual world where the senses are alive, riding along with emotionally charged scenes. Today's state-of-the-art, high-definition formats are just incredibly lucid. We almost believe that we can reach out and touch the actors. These actors, "stars" have you, enjoy unprecedented celebrity status. Whether that is a blessing or curse for themselves does not hide the fact that they are ultimately the real-life symbols of modern day idolatry (when such status is misused), which works to the detriment of society as a whole.

Directing techniques, having evolved to the standard of virtual reality, give further evidence that many do not go to theaters merely to be entertained. Moviegoers (however unaware) are more and more often viewing films to find answers to life and death questions far too difficult and embarrassing to ask among themselves, due to a pandemic form of isolationism, resulting from *hours* in front of media screens. They hunger for an "über-director" who can connect with our most inner longings for ultimate fulfillment that nothing or no one seems able to give us. The real question is—do we really want to allow these media gurus and their media to become like intimate partners in marriage and replace real human relationships? Such a detached, impersonal method of imitated reality could never do so. Still, millions allow their life decisions to be enormously influenced by the fictional characters portrayed by actors and actresses—at the whims of often highly regarded writers and directors who wish to exhibit their questionable moral positions.

Who would voluntarily allow anyone with clearly immoral and socially irresponsible worldviews (as evidenced by their products) to influence or call into question what have proven to be good, sound, morally and socially responsible principles for living? Only those who so choose to ignore them and render them somehow irrelevant! All of us are now experiencing firsthand what can best be described as an ethical slippery slope, with a mainstream media that continues to pervade as the main influence in our present postmodern society. What will we do about it? Will we accept the images and messages before us as the norm, as a progressive and politically correct worldview—or will we take a stand and speak out

> *...millions allow their life decisions to be enormously influenced by fictional characters...*

against these clearly corrupt forms of *entertainment* without compromise, in the public forums afforded to us as citizens by right? We can and should exercise such rights, and play a part in reemphasizing the importance of guarding traditional family values in order to secure the future well-being of generations to come. Our *children* are indeed the future; they are the eternal flame, a symbol of our inherent birthright as free breathing world citizens. It is what *we* choose to do *now* that will echo throughout eternity—shaping the world they will live in and influencing their choices, in turn, that they would also benefit the welfare of their own sons and daughters.

"I was not born to be forced...I will breathe after my own fashion."

~ Henry David Thoreau
On Civil Disobedience

"History is the witness of the times, the torch of truth, the life of memory, the teacher of life, the messenger of antiquity."

~ Cicero

CHAPTER FOUR

Misunderstood Beginnings: losing sight of the ideal

For just a moment, consider the latest wave of discontent toward Americans who embrace the belief that an ideal society builds on and thrives in a culture where children are brought up in a home with a nurturing mother and father. At the very least, discover for yourself that this recent rebellion against traditional family values does not represent, by any means, a new direction in culture. Throughout history, there have always been individuals who sought "alternative lifestyles" for many diverse reasons, such as the promotion of political agendas, seeking attention or self-glory, compensation for past abusive relationships, and lack of identity. Eventually you may conclude, after relevant research on the subject, that none of these so-called reasons or justifications consider the effects that such decisions have on the up-and-coming generations of whatever particular age is in focus.

Any civilization that embraced such lack of consideration for their progeny's future welfare eventually collapsed from the natural consequences of attempting to advance such selfish humanistic agendas. The Greek and Roman Empires are a few of the most well-known and obvious societies that fell victim to this social and moral compromise, which was inflicted upon the ideal cultural base of the nuclear family unit.

Today, the same agenda is advanced through the media and can be seen much like someone shouting his or her complaints angrily, hoping that by passion alone their ideology will at last be accepted. Without being fully informed on the issues at hand, however, their method of persuasion flounders. Often, their arguments sink into intimidation tactics, using verbally abusive language and accusatory rhetoric that leave

> *…Greek and Roman Empires collapsed from falling victim to their social and moral compromises…*

no room for reasonable debate, in order to resolve what is arguably one of the most important issues at hand today—to once and for all time define what is our ideal cultural identity as an American people, so there will be no confusion about what direction we are actually going as a society.

Should we not have an ideal vision of what represents the best situation and environment in which our children should be nurtured, being as they are essentially our future leaders? Absolutely! Children conceived within a lifelong committed relationship between a husband and wife are provided with the best possible role models. Such children are able to experience firsthand how their parents perfectly compliment one another in both their strengths and weaknesses, which gives them a balance of personalities. These children are provided with the best traits from each parent, so as to adopt and exemplify them in their own lives.

Again, this is the *ideal* for a model nurturing family life, which results in generations of caring and compassionate close-knit bonds among each and every member. There is no argument that today's culture reflects marital conflicts, resulting in irreconcilable differences and subsequent divorce proceedings. However, is that the *ideal* option to resolve spousal conflicts that would, in turn, affect our children's views about the importance and sanctity of the covenant bond of marriage? Absolutely not...the *ideal* is for parents to set aside their own differences and sacrifice whatever personal pride may be involved in order to reconfirm what is the most important factor within the family unit—the *children's* welfare and development. Remember then that we *aim* for the ideal, *not* throw it out and cease to pursue it because we and others tend to fall short of it. Exceptions to the rule are not evidence that the ideal family unit is fallible in itself! Rather, they point to a lack of commitment to fulfill the ideal and a willingness to allow that admitted character flaw to be passed down to the next generation. For what? Impatience? Refusal to compromise over trivial passing matters? Stubborn *pride?* It's just preposterous!

To use children as guinea pigs while society experiments with "alternative lifestyles" in order to satisfy fleeting lusts, never amounting to any lasting significance, is to thumb our noses at our own ancestors, and invite the social *and* economic collapse of our society! For anyone who questions whether these axioms hold water, it would be time wisely invested to review history for oneself. It is one thing to accept truth from someone, and quite another to come to that knowledge yourself, to make proper application of it, ultimately to benefit those in your immediate sphere of influence. What a dishonor to ignore existing historical records, which point us to egregious social trends, that we might take heed not to follow the same paths toward sociocultural destruction.

The amount of ground being gained by a minority who embrace lifestyles that do not produce natural fruit (i.e., children) through the major media market in the past few decades is astonishing. That is, considering as well what the consequence would be if *everyone* was to embrace such non-reproductive lifestyles! Is that being insensitive? Truths based on historical and indisputable *facts* have many times, I have found, offended *only* those who are unwilling to consider them in their proper contexts. In essence, they must learn from their own mistakes, often in the most grievous ways and in all stubbornness. Despite any attempts by those who have learned these lessons, in often the most oppressive fashion, and who seek to devote their lives to making every effort to help prevent others from experiencing almost certain sorrowful regret—many still refuse to even *consider* such wholesome views of family living in quite obstinate form, having already stereotyped the entire model as narrow-minded and restricting. (And remember, "tolerance" is often *their* main mantra!)

Ideally, the family model is reflected in an exclusive relationship between a man and woman as "they become one flesh" (Genesis 2:24 NASB). Our heritage as Americans cannot, then, be so easily separated from the Christian Faith from which that ideal is derived. It is *from* those practiced beliefs that our government was formed to protect the believed inalienable rights endowed by our Creator, in order that we could ensure our children would also enjoy the liberties guaranteed by the proper application of those Christian principles that bring abundant living. Once again, this is not being insensitive or offensive to other faiths. If Jesus, the founder of Christianity, teaches to love one's neighbor as oneself...to do unto others as you would have others do unto you...and our very Constitution professes and proclaims to adhere to those very principles, how could *anyone* who has enjoyed the fruits of such land of liberty be also offended by such a virtuous foundation? (John 13:34, 35 NASB)

Jesus' life is an example to follow, an *ideal.* He sought and met the needs of those who *knew* they were in need. We would also do well to be open to the needs of those around us, resulting in our appropriate response—to meet them. This *is* how Jesus* taught those around Him, and all people, to love each other and our children. In fact, if we all sought to know for ourselves Jesus' true mission—we might just find who we were always meant to be. If anything, much would be gained in becoming acquainted with this love—this giving of *self*—that is what the American family model, from the very beginning, has sought to exemplify. How can anyone condemn an entire body of such faith, for the actions of a few bad apples that corrupt its true mission to do justly and show mercy to all? Such actions reveal a clear misunderstanding of America's genuine beginnings. Shall we again examine the life-giving roots of this great nation?

"The unexamined life is not worth living."

~ Socrates

* "Jesus" is an English transliteration brought down from the original palaeo-Hebrew, ओwYaz [closely pronounced "ya-HOO-shua"], which means— aYaz is our salvation aYaz [ya-HOO-ah] is the original and proper Name given for the Creator commonly know as "GOD" or less common "Yahweh" (an attempt to pronounce the Tetragrammaton – YHWH- using English letters. Tetragrammaton=Greek for "four letters."

CHAPTER FIVE

The Spoken Word: manifested thought or the practice of subtle persuasive deceit?

Some of you may be wondering, "We're not going to harp on religion now, are we?" or, "Is this going to turn into some promotional work for some New Ageism?"

The answers are *no* and *no* (to your relief, I'm sure!). You will find the following statement, though, quite intriguing, as many I've found do:

It is impossible to fully separate religion from any aspect of our lives.

Any who would move to disagree or argue with this first need to ask themselves what it is, exactly, that they consider to be *religion*. How do you define it?

The prevailing view on religion is defined as any

practice that adheres to the principles of whatever sect one chooses (within any world belief system), thereby accepting it as a legitimate view of universal reality.

Quite *inclusive,* isn't it? The only problem is that truth itself is *exclusive,* and is compromised if such a definition of religion continues to prevail. Moral relativism ensues in such case, blurring the lines of distinction between the ethical and unethical.

Religion, by very definition, refers to a set or system of beliefs devoted to ultimate or genuine reality. True religious practice, therefore, would describe an exclusive lifestyle, congruent to an equally exclusive worldview.

The negative connotation responsible for leading so many to shy away from any discussions of the subject of religion is rooted in what others' ideas of religion are thought and believed to be. A common worldview, put forth by more than a few of the major religious philosophies, is that divine favor is *earned* through our works alone, and subsequently *buy* us eternal bliss in a place we label Heaven or paradise. By far, that is the most popular—but is that the actual or exclusive meaning fully descriptive of what characterizes religion? Absolutely not.

> "Then why would such an overwhelming majority retain such errant definitions of the word?"

It has always been shown in societies that dominant worldviews exist not because they are the most reasonable or even convenient, but because the "herd mentality"—where momentum in one direction already exists—takes less effort on one's part than other options. The "go with the flow" attitude and behavior appeals to the energy preservation instinct within us, which saves or stores excess energy for periods of extreme emergencies when we would need it (think of fat). Sort of explains widespread obesity problems in a certain way, as well. In this context, though, it's more of a mental laziness rather than a physical one.

> *...dominant worldviews exist not because they are the most reasonable or even convenient...*

Without effort to discover what religion is, in full context, one would fall into the present default view that any discussion or debate of religion is futile, since it is held (falsely, have you) that all faiths are equally valid, and an issue private in nature.

Words Fitly Spoken

Again, such herd mentalities can lead a majority to misinterpret, by taking out of context, most anything. The phrase "separation between church and state," being used to implement a secularized society, is the most obvious example of this corruption (and quite possibly the most famous).

Because our beliefs determine our thoughts, behaviors, and the *words we speak,* it is completely absurd to think religion could be so compartmentalized, and not influence governing matters carried out by those who possess them.

In a world where most (if not all) nations rely on placing enormous emphasis on the "rule of law" for keeping the peace, punishing criminal activity, give an overall standard guide of ethics, and mete out some semblance of civil justice—an important aspect of life in general (and possible divine will) is often ignored. Many times it is even completely forgotten, in our demand that our personal senses of justice and peace be promptly executed.

GRACE

The term most simply put means "undeserved merit." The spiritual dynamic has gone so unnoticed, at times, in our bloodthirsty, eye-for-an-eye world of violence, that many dismiss it as just some formal prayer to be said over Thanksgiving turkey!

Yet still, we can observe in such holiday celebrations that a *remnant* of the term *grace* still remains—the principles of gratitude and forgiveness—even though they have lost much of their true value in our society today.

Ever *give* anyone a *gift?*

Really *give* a gift?

As in, didn't expect something back in return for it?

Nothing —I'm talking not even a "thank you."

That's *grace.*

Most often you'll get the "thank you"—but you didn't *give* the gift in order to receive the thank you...or (if truly a gift) anything else, for that matter.

Think about it. Expecting something in return for a gift renders its being an actual *gift* null and void. In such a case, the object or service is actually an acknowledged loan, to be repaid at some point and time.

Giving gifts on special occasions to people we know (even if only briefly) is an excellent picture of this dynamic in action. At times such as these, it is recognized that certain recipients are "undeserving" of such gifts. Exactly. So why do we do it? At the very least, as an acknowledgment of our irreplaceable worth or value as a unique being playing an important part (even if seemingly insignificant) within the family. The same would, of course, go along with our all-important role as members of society, no matter how small or large a part we may play in its progress.

As anyone can see, after similar explanation, "gift giving" is in fact a teaching tool when interpreted correctly for what it is.

Life is a *gift.*

Did we *deserve* it? What exactly did we do to earn it? Just who do we *demand* give it to us? Kind of puts everything in proper perspective, in a way...doesn't it? (A good and effective review of just how undeserving we are at times never hurts, either!)

It goes to show that we should also speak in a *graceful* manner, as well. How often do we attempt to mete out justice with our spoken words? How often do we dismiss those who may be offering a kind word undeserving of us?

How does one best receive such undeserved gifts?

With genuinely sincere gratitude, of course.

What would determine that these ingenuous attitudes were being established in receipt of such grace? By the imparting of similar *if not greater* grace toward others in the same fashion.

...Oh, what a wonderful world *that* would be!

Not feeling appreciated? Do it *anyway*. Someone will, and that is all that matters. Think snowball effect—not shotgun blast! It will catch on...eventually. Isn't patience supposed to be a virtue? (No surprise that there's so little of it today.) Here...this is for you—though you may be undeserving—please accept it...and if you <u>truly</u> wish to return the favor—<u>go</u>, and do likewise towards another, as I have done to you.

Preventing Subtle Deceit

In our routines of living, it's not difficult for us to take the many wondrous tasks we are able to perform for granted. Abusive behaviors and/or ill-spoken words that we may have observed, or even received from others in our youths, continue to revolt us as we see similar activity persist even today. We wonder why, and even ask ourselves or others, "Was that such uncalled-for display of verbal abuse really necessary?" No, not necessary at all—just *effective*. A pure *power play*, which tears down dignity through intimidation tactics exploiting one's insecurities and fears.

We have lost our way when we allow ourselves to fall into this all-too-common pattern of corrupt behavior—out to break down the human spirit, instead of building it up so we can all live in a nurturing environment better suited to unleashing our full human potential.

It can creep in nonetheless, slowly and quite surely through the generations—most noticeably down to younger family members who must struggle with their love for a parent, yet harbor hatred and resentment for words and actions unbecoming a mother and father.

The **words** we choose to speak are powerful when used with complete sincerity. Underestimating their importance diminishes the value of language. Words often become cheapened, used merely as a tool for immediate self-gratification in this way. The corruption of language can only last for a period, though. We are known by the words we speak...Do we say whatever is necessary *just* to obtain whatever we want at the moment? Are we speaking to shift the blame to a coworker, shrinking from our own responsibility and attempting to build ourselves up for gain of recognition?

Maybe we instead take great care in our choice of words, knowing that every relationship—business or personal—hinges on well spoken and fitting phrases that reflect the often-delicate nature of the human spirit, in need of someone to take the time to empathize. Do we understand somewhat what others may be experiencing in the difficult moments in their lives? Do we care enough to bring them a word serving to strengthen them and even give them a ray of hope?

> *...our beliefs determine our thoughts, behaviors, and the words we speak...*

Such is somewhat revealing of the dynamic power of the spoken word:

"Like apples of gold in settings of silver is a word spoken in right circumstances."

~ Proverbs 25:11

Not only can our words spoken in sincerity influence those around us to do the same, they point us to their immense value in the right situations...as a master craftsman's exquisitely designed creation, made with the very best precious metals, presented to his beloved: *Priceless.*

Whether you hold the practice right or wrong, people know us and seek to discern our intents or motives by the words we use. We are completely responsible, as well, for those words we allow to flow past our lips; they are a measure of who we are and what we value, representing how much or how little consideration we have for others (not to mention ourselves).

So what place or part does the spoken word have in a progressively advanced technological world?

Words will increase in value with the increased sophistication of technology. Precision in language and communication is a necessity in the implementation of technologies *precisely* made. After all, the goal of advancing technologies is to increase efficiency, which includes reducing levels of error. In the same manner, those who demonstrate their acknowledgment that such advancement hinges on our ability to communicate in a precise and veritable way—will continue to be increasingly invaluable in our society.

> *We are known by the words we speak…*

"...out of the abundance of the heart the mouth speaks....by your words you will be justified, and by your words you will be condemned."

~ Jesus of Nazareth
Matthew 12:34, 37 (NKJV)

Reexamining language for common harmful pitfalls we should work to avoid

Given how easily words can subtly work themselves into our vocabularies (many times to our detriment), it serves us all well if we bring to light some of the more stealthy, apparently harmless, yet actually self-sabotaging words and phrases we pick up from others almost unconsciously.

Many of you will be astonished by how imperceptibly these words worked their way into your everyday language, and equally so by the "unspoken" meanings they carry.

Once I reveal the ones most commonly used, you will have been changed forever...for your eyes will be opened and unblinded to the true nature of the motives they uncover in their usage. Ready?

From this point on, pay more attention to the words others speak in your conversations, as well as the words you use. You will be sincerely amazed by how often you use the ones described below—and more importantly, why you use them:

Uh and **Um:**   not really *words*, but utterances that reveal and give clear admission to a state of unpreparedness, lack of composure, and often, meager vocabulary.

"I think":   automatically claims uncertainty (educated guess at best).

"I believe" *:   most commonly sets out to establish truth and/or implicitly reveals an individual's worldview (right or wrong).

"Good luck":   However well-intentioned, *luck* acknowledges *chance* as the governing dynamic of the universe. Why don't you just flip a coin every time you need to make a crucial decision?

"You know":   The most insidious, overused, and effective cliché used today. It can take on multiple meanings and implications. As a question, it asks if we understand or if we can relate (in the most innocent of uses), but can also practically force us into such claims through emotional demagogic appeals—especially when the speaker allows little if any time for true signs of affirmation or negation.

"You know" can proclaim that the listener automatically understands what the speaker is saying and, at its worst, program both to believe something that is, in fact, *not* true at all.

The phrase is often used throughout a conversation more than once in a single sentence which not only reveals its effectiveness but reinforces false senses of self in both the speaker and the audience.

The absolute worst aspect of the phrase is that however used there is no way to pinpoint how exactly it is meant or *intended* to be used without questioning further, or insisting that the speaker be more precise in what he or she means.

Because its use is so common and gives the appearance that we all stand on common ground on the chosen subject...it continues to be a cloaked but effective means to sway others to accept untruths and what's more in directions we would otherwise avoid. It has become moral relativity's greatest, most subtle, and sinister weapon.

"How Do We Help To Counter These Corruptions Of Language?"

Avoid using them yourself, for one...bringing them to the attention of others (tactfully) is another. They are verbal crutches that discourage effort in finding better words to explain or give detail to subjects in focus, the epitome of "lack of a better word." Just another prime example of the moral laxity of the age. Profanities are so obvious in their corruption of language they don't even rate an explanation...they are no less a confirmation of utter lack of self-respect on the part of the speaker.

So—let's say—all these years, you've been or know someone who's been deceived by your or their own use of language...feel burdened? Relax. That's what this text is for. *Forgive* yourself. Easier *said* than *done*, you say? The past is *gone*; try not to dwell on it. Use what you've learned and move on. You're better equipped now to do so—and equip others, too. Maybe you weren't ready before, if that's not too much for you to admit. If this is so...another reading of the section on First Principles (chapter 1) may serve you indispensably well. Build your life on a *sure* foundation, no matter how old you are, and you'll never regret it! No better place to start than in the present...

> *Words will increase in value with increased technology.*

"In an age of universal deceit, telling the truth is a revolutionary act."

~ George Orwell

Part two:

the SUPERSTRUCTURE

"Nothing in life is to be feared. It is only to be understood."

~ Marie "Madame" Curie

Chapter Six

Biology & Belief: rethinking the cell

The latest attempt to sterilize our language, rendering it void of any religious influence, is a brave (though futile) experiment in people management. These labors have been shown quite effectual in method—if only for a time—yet regardless of even contrived planning, none of us are able to do or say anything in all actuality without revealing our beliefs and values. We could walk around pretending to be someone whom we are not—but for how long? Society has found it much easier to suppress discussion of core beliefs in a secularized culture, and so there should be no surprise as to where we find ourselves. Does this make the main issues of life less controversial? Perhaps—but at what *cost?* Why compromise our integrity as a people?

Thomas Jefferson's well-known reference to the wall separating church and state, in his letter to the Danbury Baptist Association, is a prime example of how intended meanings can be turned on their heads to support agendas. The issue then (as should be today) was to prevent governing authorities from imposing upon the spiritual leaders of the community who gave moral direction. Today, that same reference has been the main force behind *preventing* those very moral leaders and followers from participating as representatives in a government created *by the same people for the same people!* This is at the very root of the cultural war in America. Never did the Founding Fathers intend—nor, we can be quite certain, could even imagine—a governing body completely devoid of any moral influence, yet that is just the message we are sending by not confronting such gross misuse of historical language.

What is so absurd is that it is all quite unnecessary...We do not require a complete change in direction—just a *slight* one. How does one lose sight of the principle that we do not live merely to *eat*, but rather eat *to live* and remain in good health? Yet, there are many who *have* lost such basic understanding, and quite a few who have yet to achieve it as well.

Instead of suppressing open discussions of individual beliefs and the standards for them, we should *embrace* them. More effort involved on our part? You bet; but the benefit from doing so will become obvious in time. Instead of a propped up global economy, we will have a stable, thriving one, built on open and trusting personal relations. Profit motives will come second to true personal and family needs. Is that not the environment we should be building for our children? Yes, this *is* the balanced progression of civilization we are searching for.

Progressing from the fundamental elements of social balance to its superstructures requires a substantial consensus of primary purpose. Agreement of purpose will determine overall design—since efficiency in function factors considerably into the form to be constructed for resiliency and reliability.

The American experiment in representative government continues to lead the present world in functional superstructures of civilization building. It is partisan politics (and the factions within them) that make clear it is not without flaws. They disagree over *how* government should function because primary purpose has been blurred. The term *democracy*, for example, has enjoyed a celebrated and almost sacred status—yet in purest form and in practice, it is nothing more than majority rule, or "mobocracy." Does a majority opinion determine what is morally responsible? Not anymore than personal beliefs can determine or establish what is truth. Correct beliefs, then, are an essential part of the overall

> *Instead of suppressing open discussions of individual beliefs—and standards for them, we should embrace them.*

equation in governance, and why religion cannot take a backseat in favor of some impossible, secular culture. Upon closer inspection, the implementation of a republic is clearly the better and morally responsible choice of a representative form of government.

A republic's rule relies on an order where the power of representatives is derived from a body of citizens who are *entitled* to vote them into office. What determines such entitlement is a factor requiring standardization. Giving an educated body of citizens (such as those who hold the equivalent of two master's degrees or above) the responsibility to institute these standards is an example of how best to implement a pure republic. Allowing every citizen who merely meets an age requirement voting rights—isn't even respectable baby talk. Giving these rights to those who can demonstrate acceptable levels of financial accountability, language comprehension, and other measures of logical reasoning capacity begins to reflect a more responsible, representationally governed society.

Religious liberty and the freedom to express respective beliefs are not compromised under such a form of government—but rather, are encouraged. Every human being should, *at the very least*, live among and within a society where all are given, by law, the *rights* to believe whatever they wish and to practice those beliefs so long as no violence (less self-defense) comes from those actions towards others. Liberty *and* justice for all. Such an inherent right to life has become clouded in a society immersed in a nonstop informational media *blitzkrieg*. Filtering out all the irrelevant (and often harmful) data is much like trying to focus on one specific radio station while the others are tuned to different stations playing at the same volume. Again, difficult—but not altogether impossible.

Tyrannical forms of government can not long endure. It is simply impossible for the free human spirit to be fully expressed and observed in a restricting or oppressed environment, and not lead to an eventual deliverance from such ill conceived bonds. Through open dialogue, thanks to freedom of speech, historical revisionism can be confronted and essential truths can once again be made known to set our feet back on the path towards a more perfect union, which thrives on individual contributions and accomplishments that raise the standard of living for everyone. Albeit, much of this depends on our ability to contain technology as a tool towards that goal—

otherwise, we can easily be and are very capable of becoming enslaved by its innovative systems. Such containment, however, is and will continue to be a challenge. The next chapters will lend insight into such a possible (if not inevitable) future.

For now—a look into biology (in connection to beliefs) will serve as an excellent primer to reaffirm our primary purpose for existence, both on an individual and corporate basis.

***

Henry Ford offers us rich insight with his statement: "Whether you think you can—or you can't...you are correct."

There should be no question—our beliefs determine our worldviews. How we see the world and all that goes on within it *directly* influences how we choose to interact with it. Right or correct beliefs, therefore, are crucial—as they affect *every* aspect of our lives, from cradle to grave.

Ford has shown us that there are those who take a "can do" or optimistic attitude, and those who choose the "can't be done" or pessimistic attitude. Those who take the former at least make an attempt, despite the obstacles—with determination and the right or efficient application of skills, there is a possibility to achieve success. With the latter, there is no possibility, since there is no attempt—no effort is put forth—and so the outcome reflects the attitude.

> *How we see the world and all that goes on within it directly influences how we choose to interact with it.*

Acting on wrong beliefs has its consequences, as does acting on right ones. Hopefully, through or with aid of this work you will seek and discover the source of truth and a standard for morals for yourself. In doing so—by putting into practice those right beliefs—you will find that the resulting consequences are shaping and molding your character, instead of acting to punish you for breaking civil, criminal, or spiritual laws inherent to us as created beings.

Cause and effect, trial and error—these are our most obvious and revealing learning devices. At times the latter can be a lengthy and costly process, but for some it is the only passage to achieve success. We *can* learn from others' mistakes—either from those around us or even from well-known and documented historical figures—*or* we can learn the same lessons for *ourselves* without any guidance whatsoever, if we so choose.

The former, of course, is recommended. The latter, in this day and age, is considered somewhat foolish considering all the sources of wisdom available through internet search engines—not to mention having the ability to reach the actual authorities of those sources via e-mail in an instant.

The New Science: Epigenetics

Molecular/cellular biologists offer extraordinary new insight into the most efficient systems of organized community. After all, as human beings we are comprised of cells by the millions. They are our building blocks...yet each cell is specialized in function. Each has a particular task to carry out, allowing the whole system of the "being" to perform as designed. Groups of cells form tissues; those communities form organs; those organs form systems; and those systems carry out bodily functions. It is the ability to carry out these functions that makes us able to accomplish highly specific purposes or specialized tasks.

What today's cellular biologists are helping us see is how relevant individual cell behavior is to our own lives. The cell itself is a tiny community from within—even more complex than we once thought. The cell, it is found, has *intelligence*. We can even say it possesses a brain to carry out intelligent functions like interacting and adapting to its environment...it acts rather *human*!

Well, it *is* human. *We* are, in fact, functioning communities of cells, all working together to accomplish specifically chosen tasks. In quite inspiring terms, we are composed of *several* different communities, living separately yet all equally important in function, in order to accomplish a higher purpose than could be achieved individually.

That should definitely hit a nerve!

Where we differ as a civilization is that individual persons or beings have lost this type of communal spirit, and choose to be cut off from this kind of higher purpose functionality...not because of simple or absolute rebellion alone, but due to individual and corporate bodies choosing to seek fulfillment outside of this efficient means.

The advent of the technically innovative internet,

however, is able to connect every individual and corporate entity, much as each cell connects via the nervous system, which includes the brain.

The Brain vs. Genetic Determination

For quite some time, academia has led us to believe that genes, composed of DNA strands, control our biology (or determine our behavior). The brain of the cell was also widely assumed to be located within the cell's nucleus. This faulty belief was found to be without merit when an experiment was performed by removing the nucleus from the cell (enucleation).

Remove the brain and, it was said, the cell would cease to function and die—yet, the enucleated cell *continued* to function! What it is unable to do, however, is reproduce itself or replace defective (or worn out) protein parts.

This was the clear evidence needed to conclude that the nucleus and genes of cells contain the vital information for reproductive purposes (making it the gonad). The nucleus is our *hard drive*—it is where our information (our blueprint) is retrieved from, stored, and duplicated.

Think about it...our brain is responsible for controlling our *behavior*—it's attached to our nervous system, our *senses.* Our senses are directly exposed to our environment—that's how we interact in and with it.

So, where is the brain of the *cell?* What part is exposed to its environment? The *membrane* is. It is through the outer shell of the cell that it interacts with its outside world. The membrane identifies nutrients, allowing them in for digestion for the cell's energy needs, and expels waste through the same barrier. This is equivalent to our own brain's processing of information through our senses to determine what foods can be used for energy, and which ones should be rejected and/or expelled from the body.

The similarity of function from a single cell, to the human being, and then to a whole society is evident. What is less evident is the actual consummate goal of these processes as they surely strive to transcend one another from lower to higher states of being.

Solving the Culture War

What is achieved through the organized efforts of individual parts for the sake or benefit of the whole is *Increased Awareness*.

We have an innate desire to be as aware of our environment as possible. This is both for the sake of our survival as well as for the sheer joy and curiosity of exploring our vast universe. As a people, we have never had the opportunity to be more aware of our environment on a global level than we do at present. We have also developed highly efficient divisions of labor— just as multicellular communities that make up our bodies possess—specializing in functions to benefit the whole, which eases the burden on each contributor.

The current events specialists we call "news

correspondents," working in direct conjunction with World Wide Web and live satellite feeds, bring those of us linked to these media near instantaneous awareness of world events (provided this news is objective and void of "spin" or bias towards power hungry agendas). Being at a point in time when optimum awareness allows us (as a global community) to super-specialize our divisions of labor, this unprecedented access to information allows us to refine our systems for ultra-efficient management, which in turn helps us achieve never-before-experienced levels of personal prosperity, easily afforded to anyone who so desires an increased standard of living.

It is, has been, and always will be a never-ending process—with the near full responsibility for continued success falling on our shoulders.

"Be fruitful and multiply, and fill the earth, and **subdue** (manage/control) *it; and rule over the fish of the sea and over the birds of the sky and over every living thing that moves on the earth."*

~ Genesis 1:28 (NASB)
(emphasis and parenthetical added)

We can easily see (even without spiritual guidance) that we have the *power* to manage the environment we find ourselves in (regardless of scale) through the *actions* we choose to implement. Personal responsibility has, therefore, always taken precedence in the determination of the desired outcomes of intended goals.

Actual results can be seen in an increase in efficiency towards an unknown end, or adversely in a breakdown of societal structure due to loss of focus, and eventual loss of original purpose...an essential failure of central intelligence or "brain function."

As discussed:

> Remove the brain, and the organism dies in short order (having lost its ability to reproduce). Remove the brain and suffer a loss of purpose. Remove purpose for being, and corruption from within ensues.

Self-destructive human behavior is also not without its influence and consequences, being as the same is derived primarily from acting on wrong beliefs.

Beliefs, therefore, not only *control* biology, but sustain and determine its health/prosperity.

Sugar pills causing the placebo effect in volunteer patients in drug trials are the best example of this "faith to biology" connection. Patients being administered a placebo, (actually *believing* they took the genuine medication) prove that our "faith" or beliefs have direct effects on our bodily health—again, individually *and* corporately.

Aren't we seeking to increase our awareness on both levels? Absolutely. Then we have certainly retained our purpose. That's a good sign! All is *not* lost—so, doomsayers, be gone...

Yet, we can still see present in others the

symptoms of self-destruction through aberrant behavior, which acts on society as cancer cells destroy a body from within. Self-indulgent, superfluous lifestyles are clear evidence of the growing presence of such corporate cancer. A decadent society that returns to personal responsibility, instead of one that supports its citizens who play the victim, will successfully remove these cancers from within. It is evident, therefore, that we each take on great significance in our relationship to society as a whole—through our actions—stemming from our core beliefs and values.

Beliefs *do* then control biology...and the best aspect to remember about this essential truth is that we all can each take moments of personal inventory of our beliefs, and determine if they are truly *correct* ones. If not, we can *change* them and reap *all* the benefits from doing so...physically and spiritually! No matter how long you have believed or have been led to believe otherwise, no one is ever locked into making choices that turn out to be imprudent ones after further consideration or personal introspection. Make that change! Throw off the chains of tradition holding back your full potential to not just live but *thrive!*

> *...we take on great significance in our relationship to society—through our actions— stemming from our core beliefs and values.*

Inspecting our foundations for belief can easily add *years*—even *decades*—to our lives, once we lay aside urban myths resulting in destructive behaviors and lifestyles to finally grab hold of essential and timeless truths that benefit ourselves and those around us.

Once begun, this personal inventory results in a gradual tuning in and attaining of higher levels of corporate self-awareness, as well. A sense of *holistic* healing will become apparent as we bring our behaviors/actions in line with our original design—we begin to function as intended...with *purpose* and *passion*.

We have *always* had a say in our destiny...there has always been a *choice.* The same goes for any choice, for that matter—whether it is personal, cultural, national, or global.

This similarity among parts of a whole, this micro-macrocosmic relationship, is seen *universally*...just as the drive towards increased self-awareness is universally present, from the most minute life-forms to the more complex ones known to us today.

These "self-similar" repeating and growing patterns of being are referred to as a type of *fractal evolution*, which can be observed from atoms to galaxies.

Your beliefs become your thoughts...
Your thoughts become your words...
Your words become your actions...
Your actions become your habits...
Your habits become your values...
Your values become your destiny.

~ Mahatma Gandhi

"Science without religion is lame—Religion without science is blind."

~ Albert Einstein

CHAPTER SEVEN

A Fractal Universe: from atom to galaxy

The repeating yet irregular patterns that can be observed in nature's life systems have been difficult to describe in layman's terms, much less illustrate, until recent years.

Advances in geometric study have changed this illusive realm for us. We've discussed how different elements of our own bodies relate to one another as they function as a whole. The field of geometry is similar in the respect that its focus is on assessing the way different parts fit together in relation to each other. In the natural realm, however, basic geometry—with general mathematical formulas to describe simple, easily graphed patterns and structures—falls short. The reason is that in nature, irregular patterns of structure and forms are the norm, and even surpass the three-dimensional realm on many occasions.

"Fractal" geometry, discovered by French mathematician Benoit Mandelbrot in 1975, is able to more precisely illustrate these irregular patterns we find in nature. The "Mandelbrot Set" shows the simple origin of such patterns, which appear infinitely complex when observed in natural form.

The Mandelbrot Set formula:

formula 7.1 $n_1^2 + n_1 = n_2^2 + n_2 = n_3^2 + n_3...$

Actually placing a number for the variable will bring clarity to a simple origin, the same repeating formula, to an exponential increase in values—which give the appearance of a fabulously complex structure when graphically represented using computer illustrated models (*see <fractalwisdom.com> to view various perspectives of these representations*).

formula 7.2 $1^2 + 1 = 2^2 + 2 = 6^2 + 6 = 42^2 + 42 = 1860^2 + 1860 = 3,263,442^2 + 3,263,442...$

The formula remains the same, even though each resulting product takes on drastically increased value at an exponential rate.

Pick up a leaf from beneath a tree...notice its veins like tributaries of a river. Now observe the twig on the tree from which the leaf fell...the branches from whence the twig came...and the trunk where all branches are attached. At each level, you can see the same basic pattern form as the result of the process of growth. Yet, looking at the tree as a whole from a distance renders the self-similar patterns seen up close less obvious.

When we look at each other, we don't see a community of specialized groups of cells, but we are aware that *that* is what we are ultimately composed of.

We can, upon closer observation, see similar patterns from the nano level to the galactic...

Electrons orbit a proton/neutron nucleus, forming an atom; moons orbit planets; planets orbit a common sun (or star), forming a solar system, which resembles the spiral galaxy of which it is part (ours being the Milky Way galaxy).

At the galactic level, we see a dramatic change in value as in the Mandelbrot Set's fifth level. Formula 7.2 shows a jump from a two digit to four digit integer, and an even greater jump from four to *seven*—from a thousand value to a several million value—at only the sixth stage of the formula set. Though the scale of value becomes enormous and seemingly indistinguishable from the original or first state, the formula remains identical.

At the galactic level, the same dynamic takes place. Not all galaxies are spiral ones, though difficult to imagine. At such a massive scale, form becomes indistinguishable from its common, more familiar patterns of constituent components when observed at lower levels

From our present being, we are unable to define exactly what form and functionality the body of galaxies take on, due to the enormity of such a level.

What is of interest, though, is that our understanding and near mastery of matter's most fundamental constituents will eventually lead to the vehicle able to bring understanding also of what is now incomprehensible, regardless of scale.

Building or manufacturing at the atomic level, or "nanotechnology," opens great hopes in finally ascending to this next level of awareness. It is the next logical stage in the fractal evolutionary ladder.

Materials made from such atomic/subatomic particles will bring greater efficiency, predictability, and strength to the structures made from them, which will eventually lead to even further transformation of human environments.

Structures will seemingly construct themselves—after all...nano-sized robots will be invisible to the naked eye (being one billionth of a meter), while building (or destroying) according to program and/or project design parameters.

Yet, if we forget or fail to know the cultures from which we evolved—having arrived at this point of technological progression—will this new manufactured state of reality serve us to achieve our desired goal of a greater state of awareness? Or will we end up serving the technology, and lose our humanity in the process?

These are the challenges facing us today. In the meantime, as these technologies begin to emerge and become realities, it serves us well to develop the sociocultural aspect of our humanity. Without such development, we will surely become a population in the billions, with each person possessing their own version of reality without any perception of universal truth. A completely divided, self-sabotaging global culture will be the death of us all, bringing about the destruction of all our natural resources and, hence, the Earth itself.

Can you imagine an entire population numbering the hundreds of billions, living amongst each other in a perpetual state of harmony and fulfillment of purpose?

This community exists *today!*

It's called a healthy human body, living in accordance with true purpose by design. We, too, can function as such, if we learn from the societies within us. For us right now, it begins with being aware and helping others become aware of our relation to the whole— where we have come from, who we are, where we are headed now, and where we should be headed.

This is living with universal purpose towards a greater awareness, ascendancy that is more in tune with holistic existence able to sustain even greater populations towards the same.

"Look not for signs to be observed for it is not 'here' nor 'there' as some say it is...the kingdom of your Creator is within you."

~ Jesus of Nazareth
Luke 17:21 (paraphrase)

CHAPTER EIGHT

Nanotechnology & Beyond: the next frontier

Space may still be considered the final frontier...yet in order to explore the reaches of the seemingly infinite universe, we continue to discover how harnessing the power of the most minute particles of matter can better get us there. As astronomers keep an eye on galactic star systems and for signs of life, nanotechnologists are focused on how we can manipulate or rearrange atoms to build systems that not only mimic nature's efficiencies, but also reveal possibilities of actually improving on it.

It comes as no surprise that the term nano has become a buzzword of the age. How small is this sub-micro measurement? To gain easy and quite proper perspective, a nanometer is to an inch as an inch is to 400 miles. About as simple as it gets. Technically it is defined as one billionth of a meter, and is roughly equal to the diameter of five carbon atoms placed end-to-end. It is not difficult to assume that matter could not possibly be broken down any further than this atomic level, an easy mistake to make for those without access to the astronomically-priced atomic force microscopes (AFMs) used within the field. Those that use them know better. We know now that the smaller we go, the more we see matter existing on even smaller scales. Though foundational works such as Eric Drexler's Engines of Creation (1986) on nanotechnology are barely emerging on these smaller scales, we know that as we master the nanoscale level we'll be able to examine much more than mere glimpses of them shortly after.

To satisfy the curious, picotech will emerge as the next engineering scale taking place on the subatomic level (think electrons), at one trillionth of a meter. Femtotech will follow working on the quark level, or one quadrillionth of a meter. The level of technical/engineering efficiency that these scales may prove to offer is mind-boggling.

Ponder that nanotube transistors will operate at speeds measured in Terahertz (THz), or a trillion cycles per second. One cubic inch of such circuitry will be 100 million times more powerful in computing data than the human brain.

Attempting to Improve Upon Nature

Having once observed a universal pattern for life, a sensible step to better understand our own lives would be to reverse engineer the smallest self-replicating component of our being. Seeing that we begin as a single cell, eventually becoming a mere extension of it, we would obviously have much to gain, learning about its design and subsequent modus operandi. A prevailing theory is that by gaining such insight, we may be able to even improve or enhance biological systems by implementing nanotechnology.

We know now, as discussed, that matter is composed of even smaller particles/components than atoms, which is the subject of the emerging field of quantum mechanics...yet for now, we have just begun to build (for practical purposes) at the atomic level. We'll surely move on to subatomic engineering, but even then we may quite possibly find even more than quarks to deal with!

As it stands, nanotechnology offers more than enough far-reaching possibilities for practical application than we can even imagine...Here we'll look at a few, while discussing some major obstacles barring mastery of the promising mechanics.

Self-Assembling Systems of Life

In the cell, we observe a model where an encoded design (from DNA) is copied (by RNA strands) and whose inherent program is then implemented through its ribosomes. The life system is carried out in an orderly manner, according to the DNA blueprint.

Essentially, ribosomes are organism-building bio-machines producing on the molecular level—assembling atoms from a specific set of instructions.

Through reverse engineering, we have begun to

imitate such self-assembling behaviors by building nanobots. For nanobots to be practical, self-assembly (and self-replication, for that matter) is essential. Building such pseudo-biological machines individually is absurd. Let them build themselves—as our cells do.

The nanobot, being fully autonomous yet measured in mere atoms in size, is able to carry out tasks according to a desired program—offering an unprecedented number of practical solutions to address the inefficiencies of the human condition. Not only does each possess the capacity to sustain itself, by seeking out its energy needs—in mass swarms, nanobots can construct (or deconstruct) matter at speeds and scales quite impressive by any standard.

Climate-controlled brick-and-mortar buildings will be completely unnecessary since environments can be instantly brought into existence by little more than thinking them into being.

Carbon nanotubes (CNTs) give insight into the superior quality and performance of the materials to be used by these atom-sized, ultra efficient machines. CNTs are nine times stronger than steel, yet an acre's worth of the fabric weighs a mere four ounces. Conductivity is also superior...being a thousand times more efficient than copper for electrical current, and able to withstand temperatures up to 840°F. Infinite structural complexity capabilities for such super-materials allows for just about any substance imaginable to be formed atom by atom, to the specifications of not only engineers or designers, but of you, me, or anyone, for that matter.

Nanomaterials are one thing—nanoelectronics are another. For nanoscale circuitry to be considered reliable and error proof, self-assembly is crucial. Manually connecting circuits using conventional materials is much like trying to connect two dump trucks to the ends of a pencil. Circuitry on the atomic scale offers difficulty in architectural integrity, given the delicate nature of frequently occurring individual atomic bonds. To improve resiliency and reliability, laying one set of parallel circuits over another in a perpendicular fashion offers electrical signals a redundant grid on which to travel. This type of crisscross weaving overcomes areas where individual circuits are compromised or broken. Signals just move to the next available intersection—obviously with minimal rerouting taking place. At most, signals would travel in zigzag fashion, losing little point-to-point speed.

The Killer App

Again, we stand on the cutting edge of technological innovation, and the possibilities seem endless in transforming ourselves and our environments into more inviting places to inhabit.

The immediate sociocultural consequences are huge. How will the general public react to the transformation of urban and rural landscapes, being that, in reality, such traditional construction techniques will become obsolete? Labor-intensive projects become quite unnecessary, and prove to be an inefficient use of human resources in a world where molecule-sized nanomachines are able to produce desired environments at our whimsical bidding.

Energy needs can be abundantly met by solar

power, harnessed by hoards of nanobots carrying on a plethora of other tasks simultaneously. The impact of nanotechnology on this and other traditional business sectors is of no small concern to those whose lives and families depend on their current means of support. As a result, a gradual phasing–in stage is necessary before this emerging age of molecular engineered reality is to set in.

And set in it will...according to Ray Kurzweil, a leading advocate and visionary of artificial intelligence (AI) and bio-tech convergence, nanobots introduced into our bloodstream remains the "killer app[lication]." By destroying cancer cells, removing debris, or even correcting errors in DNA sequencing—disease and the aging process will have been effectively reversed. They are quite literally what constitute the genuine fountain of youth.

Is this the coaxing innovation that makes life extension for humans of a few hundred years or more a common expectation? That remains to be seen, of course, but is not a stretch of the imagination by any means. In the meantime, nanotechnologists such as MIT's Angela Belcher, a specialist in biomimicry, will lead the way in improving upon nature's already efficient means of growth and sustainability. A profound transformation of business, commerce, and overall society will soon follow as a result.

"Why does this magnificent applied science, which saves work and makes life easier, bring us little happiness? The simple answer is because we have not yet learned to make sensible use of it."

~ Albert Einstein

"I don't really think about it, I just do it. In my mind, my hand is still there."

~ Jesse Sullivan,
on using his brain-powered bionic arms.

CHAPTER NINE

Cyborgenics: taking the techno-convergence seriously

The dynamically changing face of humanity can go nearly unnoticed without a constant awareness of the sociocultural metamorphosis taking place. Reflecting on unprecedented global transformation in societies as a result of technological invention and innovation serves to maintain this vigilant cognizance. The speed and subtlety with which these changes are progressing requires it. If not, we would surely take these vast improvements for granted and begin to consider them entitlements.

The personal conveniences afforded us were not intended merely to raise our quality of living, but also to give us more quality time for endeavors that benefit ourselves, our families, and those around us. As human labor is increasingly replaced by machines through automation—robots show more ability to handle service-related functions, and computers continue to become exponentially efficient year after year—our role in society (as we have known it) will either change dramatically or cease to exist altogether.

Work, for many, does not carry the same connotation for today's postmodern citizen as it has traditionally been regarded. Labor has been translated from an arduous, physically demanding chore to management of highly diverse and specialized robotics and computers common in today's workplace. The intellect has displaced brawn, making it necessary to develop technical knowledge and skills to rightly implement this new automated workforce.

A more valuable manager-worker must emerge in this new and improved, globally connected environment if we are to remain indispensable members and benefactors of society. It is by no means a stretch, then, to realize with our direction— from the internet, to wireless

> *A more valuable manager-worker must emerge...*

peripherals, to smart pills, to nanotech—that an actual integration of digital devices with the human body is one that is already underway.

The Beginning: human enhancement/ digital thoughts

Science fiction it is no longer. The future we once dreamed is present reality. The technology is among us, progressing swiftly, and has brought the birth of a new race—the "cyborgs" (cybernetic organisms); more specifically, digitally and mechanically enhanced human beings.

Very few have faced the utter hopelessness, discouragement, and frustrations that paralytics experience. There is no question of the present genuine desire of individuals who wish to bring relief and hope to this portion of the disabled population. Such is being accomplished and is in the refinement stages as we speak, through the latest inventions in biotechnology.

Matthew Nagle, a victim of neck-down paralysis, marks the birth of the true cyborg. Three years after a brutal knife attack, Nagle, 24, agreed to undergo implantation of a device for the first human trials of BrainGate. The brainchild of Professor John Donoghue, a Brown University neuroscientist, BrainGate translates thoughts into practical application, allowing the user to move a cursor and select icons with mere thoughts alone.

Late 2004 marked initial success of the science after Nagle's three week recovery from implant surgery. BrainGate is composed of 96 hair-thin electrodes, each a millimeter long. An implant about the size of an aspirin is pressed unto the brain's surface just above the region known as the sensory motor cortex, which governs arm and hand movement. A wire connects to the array, extends from the skull to an amplifier, and finally to a computer via cable. The brain activity is deciphered and then translated into useful actions.

The capturing of electrical signals generated by thoughts and the ability to decipher them was just the beginning of BrainGate's success. Developing the software to translate the information into increasingly practical application was the next and final breakthrough, which proved the technology as an extraordinary discovery and new hope for paralytic men and women. Even the most basic functions of cursor movement and selection of icons allows one to send and receive e-mails, turn appliances on and off, and control robotic arms to move objects.

These fundamental functions allow a once fully dependent population to become an employable, semi-independent workforce. To give a person such a sense of self-worth must be a tremendous sense of accomplishment for Professor Donoghue. For those on the receiving end, who are for once able to make a meaningful contribution to their society and fellow man, he must seem a miracle worker. The self-employment opportunities are endless.

History will mark BrainGate as the successful merging of man and machine. Still, BrainGate was impractical, invasive, and aesthetically unappealing, considering the implant was in direct contact with brain tissue. Such intrusion penetrating the brain caused some minor tissue inflammation.

A Step Closer...

John Wolpaw of New York State Health Department's laboratory arm, the Wadsworth Center, proved that brain signals could be captured with electrodes worn directly on the scalp. By wearing a cap not unlike those worn by swimmers, the case that one needed a direct implant has been proved unsubstantiated and unnecessary to capture signals.

The Flip Side

Enormous possibilities will open up for the disabled through this technology—that much is obvious.

The independence gained or regained will continue to be miraculous in itself, if only recognized for the greater amount of human dignity given to the paralytic population. What can also be seen, however, is the ethical question as to those who would desire to apply the technology to fully functioning human beings to *enhance* their humanity—to become *more* than human. Such a being would not be *human* any longer in the technical sense. A redefining of *being* had to take place after all.

Thus the emergence of an entirely new race has begun—cyborgs entering populations pose difficult and complex ethical questions:

- Just how would human rights apply to cyborgs?

- Could they at all?

- Would such even be classified as humans?

- Would they become a slave class, or would mere humans become slaves?

That's just a start. The list is proverbially endless.

Enhancements in memory and thought processing would render those opposed to human improvement—"obsolete", as the case is made. Left in our natural state, we would simply not be able to keep pace with the artificial intelligence, or AI, products that would be made available to us for outpatient implantation. Would such a "once human, now cyborg" retain true compassion and emotion? That aspect is highly questionable, as priorities and values in efficiency have already been established and may override maternal/paternal instincts in the long run. Would they be *more* or *less* trustworthy if they did?

As you can imagine, the questions become increasingly complex and often paradoxical in nature. The acceleration of technological change to exponential rates beyond human comprehension or control is referred to as the "AI Paradox."

As human and cyborg systems begin to interact, and technologies become even more complex they—in a sense—become totalitarian. How so? If the technologies do, in fact, grow more complex than humans can comprehend, there will exist no one competent enough to oversee or supervise their use or application! Therefore, humanity becomes completely subordinate to their systems, as the argument goes.

"Technologies" have always been present, to a degree. The individuals who understand the latest technological innovations naturally have some control over the societies that thrive and/or survive by them. But what occurs when the technology surpasses human comprehension? Who's in control *then*?

The argument of "transhumanists," who long for and foresee a transcendence of humanity into posthumanity, is that, in order to remain in control of these ultra-complex systems, we must *ourselves* merge with it! The study of this integration has emerged as the field of "cyborgenics".

A main term discussed in the field is *singularity*. Singularity is described as the event when all human consciousness is unified into one *singular* mass consciousness working in unison towards a supreme will. Before this singularity occurs, an irresistible force absolutely recognizing human transcendence for physical survival will become increasingly apparent.

Because the human mind will not possess the ability to follow, comprehend, or even control such systems, the implication of further progress can no longer be foreseen since new elements and intentions enter into the equation far too rapidly to anticipate.

The only way conceivably seen to avoid this quagmire is to increase our evolutionary pace. We must (by necessity and survival) become more than what we are. There will be those, of course, who will by choice refuse human transcendence for moral reasons. So be it— it is a choice that will have to be made. Physical survival as an instrumental mechanism will overwhelmingly be the choice of the many, as a rule. As a result, posthumanity will eventually overcome human existence, and most likely the reason (altruistic or otherwise) for choosing to become a cyborg ceases to even be relevant!

Not quite the future you imagined? It's not as far off as you think. No need to fear—just be aware...be very aware.

In the right context, cyborgenics will serve as one of the more useful innovations known to man. As can be seen, however, keeping things in their proper contexts is not one of humanity's strongest abilities...

"...this is not an alien invasion of intelligent machines coming to compete with us. It is already in our pockets, will soon migrate into our clothing, and will make its way into our bodies and brains."

~ Ray Kurzweil

"Nothing is so common but the wish to be remarkable."

~ William Shakespeare

Chapter Ten

Chasing Immortality: heavenly bliss or endless hell?

The hope for scientific breakthroughs to yield a genuine fountain of youth has never been closer to fulfillment. The form which that eternal spring will take remains speculative and ever changing, as technological innovation increases. Will it begin with anti-aging drugs and cancer, or virus killing nanobot warriors deployed into our bloodstreams? Will it progress to the manufacture of organic tissues replacing aging body parts and altering of genetic structures from the point of conception? We have already proven the above methods possible...the technology is here and steadily improving, exponentially.

Just how far are we willing to go in this pursuit to stave off the effects of aging? Is it truly an inevitable fact that we must face an event such as a singularity? Once a few justify using synthetically made replacement body parts, not out of necessity, but out of superiority of the material alone—have we not passed the point of no return to face the extinction of humanity?

The science fiction visions of stainless steel and titanium bodied "Terminators" now seem rather absurd and impractical, really. Carbon fiber sounds more likely the material of choice for its strength, durability, incorruptible nature, and inert properties to start, then there's always its light weight: perfect for creating tomorrow's *übermensch*, or superman.

The search to build a better human, as we have discussed, may have grave consequences if certain ethical limits are not maintained *and* enforced—the worst (or best for some) being the end of humanity itself, and the domination of an entirely new race of cyborgs solely dedicated to economic and energy efficiency. A pure secular soul will have emerged then, void of any emotional response or moral conscience.

Is it possible that this search for eternal youth is simply a product of the misperception of death or mortality itself?

All around us, we see the image of a Creator through the various life systems present. The fractal relationships have been clearly illustrated. A seed dying in its form to become a seedling, which more resembles its parent plant is one excellent example. The death, however, was just in *form*. The life of the organism never ceased in its being, but *transformed* into a higher state when nurturing environmental conditions were met. Life continues on, just in a different, more suitable, and highly efficient form.

Our perceptions of death are often rooted in tangibles. We view death as an end—a ceasing of existence—but is it? Surely, we live on through what we leave behind—our personal influence on and through...children, artistic expressions, humanitarian trusts, etc. It is a cessation of consciousness that we wonder about. Do we continue to have thoughts, as we do when we are present in carnal bodies?

There are still enigmas science has yet to explain on such subjects. We can, however, through simple evidence in ourselves and the rest of nature, place hope in the fact that life is often *transformed* into higher states of being at certain stages in the lifecycle. It is not, then, a stretch to believe that physical

> *We view death as an end—a ceasing of existence—but is it?*

death is also a stage of transformation into a type of higher (and even more spiritual) state of existence or being.

So, are physical death and aging things to be avoided at all costs? With a life lived with a pure conscience towards others, seeking to benefit mankind to the best of one's ability, it would be a time best suited to reflect on a fulfilled life with full certainty that one is prepared for whatever the future holds in an afterlife. I cannot fathom a more heavenly, joyful, "eyes wide open" way to leave this physical realm.

On the other hand, I cannot imagine a more miserable, hopeless, and fearful way to leave, having lived a life selfishly, looking out for one's own interest alone, and at much expense of others—purely egocentric in nature.

To live with a clear conscience *or* face fear of the unknown...

Not a hard choice for some. For others, the question of uncertainty and absence of tangible evidence pointing to the existence of an afterlife is, at present, an insurmountable obstacle. So willing are they to practically forfeit their souls for a mere lack of effort in striving to answer the unknowns. So many have given up in exchange for temporal sensual lives of personal creature comforts that are able to soothe, medicate, and silence their failure to cross the finish line, which they must face as their own undoing.

Wagering personal assets on the possible outcome of a contest is one thing...betting one's possible *eternal soul* on lack of scientific evidence is quite another! Science has come a long way, but has much further to go. We have just *begun* to prove genetic determinism as an unjustified theory. Discovery is a process—for science and for us, ultimately, as individuals. We all have a personal responsibility to contribute to that ongoing process or we have lost our individual purpose toward a common good and have become cancerous: a spreading influence intent (even if unconsciously) on sabotaging the functional whole, which is at the moment making great strides in new scientific discoveries. Unprecedented!

The most obvious cancers of society are terrorists whose fundamental beliefs leave them no other option but to act as suicide bombers—taking with them those who oppose their beliefs. Why do they have no other option? Because within one major religion, there is no guarantee of a heavenly afterlife outside of destroying those who do not hold their beliefs as dictated by their god—a god they believe is also merciful and also gracious!

How twisted can such understanding of the divine become? To define *mercy* by killing, through violence, the very creation their god claims to have made—in order to secure a place in a blissful afterlife? The depth of self-deception that the desperate human soul reaches is further than most of us would care to explore.

To come full circle, we must confront those living out self-defeating lives and attitudes, not just with our *words* but through our own *lives*. By basing our beliefs and subsequent actions on *truth* and not *determining* truth by our individual beliefs—we will then live by demonstration and prove our unwavering hope and joy with confident assurance. Use of violence to convert, by force, those possessing opposing beliefs should never and need never even be considered a respectable option! We will then be prepared to give answers to those who ask us for the confident hope that we have…

- We are then informed on the issues that matter and can explain why applying first principles is not only relevant but essential to make responsible choices and rendering just decisions.

- We are free from the guilt and shame of our past failures to move on and strive to succeed in present and future endeavors.

- We are immortal already in what we leave behind after we have left our physical bodies—at the very least—and the possible transformation, into a higher state of being leaves even greater hopes surpassing our understanding.

Drinking of the proverbial fountain of youth, through the latest technologies, has reached us sooner than many thought possible. Make no mistake, for those who drink now, you are the test subjects—the guinea pigs. The progression towards this "perfecting" of science will be so subtle and incremental that there will be no need to entice you with bonus incentives or privileges… you'll just follow the masses—go with the flow—participate in the latest socially acceptable trend. The great experiment in socialism—which has yet to ever succeed—always tends to creep its way into promising social justice for all, while secretly implementing the Hegelian dialectical method/agenda (*see Glossary*) of an elitist class. Oh, you won't even notice any difference—slowly and surely, humanity will cease to exist without resistance. (You can barely unglue people from their flat TV and computer screens as it is *now!*)

So, what will *you* leave behind?

Have you made a difference, however big or small? To be *truly* remarkable, answer the difficult questions posed here for yourself…not to be seen or noticed by others, but to be fully prepared, come what may.

In this way, you'll know peace, and be in a position to help bring it to those around you amidst times of personal grief or cataclysmic disaster.

After living by right beliefs and enjoying the fruits thereof…chasing immortality will begin to seem like a dog chasing its tail! Humorous to watch at first—tragically pointless and self-defeating, once realized.

"Wherever your life ends, it is all there. The advantage of living is not measured by length, but by use; some men have lived long, and lived little; attend to it while you are in it. It lies in your will, not in the number of years, for you have lived enough."

~ Montaigne

"An individual takes on significance only in his/her relationship to society as a whole."

~ Brian Herbert

CONCLUSION

Where do we go from here?

The culture war is all about the meaning of humanity—the meaning of life itself. Moral relativistic worldviews leave no lasting solution, though they may outwardly bring a semblance of peace, at present, in an evolving multicultural global society. Polite conversations and brief oral exchanges, which mask hidden motives and agendas to convert others over to major world religions, are no less deviant behaviors than any other. Yet, the practice of these lifestyles remains utterly pervasive and encouraged within our society through the popular media.

For those who identify with the overwhelming factors involved—take heart...

The obstacles are not as insurmountable as they may seem to appear. In fact, your endurance through the tragic events we have faced, has in reality *prepared* us to bring others out of the waste products of this culture. Those products include persuading individuals that they are hopeless victims to their genetics, their family heritage, or their poverty; that they have absolutely no control over their destinies; that our fates are already decided, and that there is no use in even attempting to make a positive difference anymore or at all—seeing such acts as utterly futile.

We began with first principles—and this is where we must begin in all matters of international affairs, foreign or domestic. Virtual meetings via the internet, cable, and telecom services have made international borders irrelevant in the business world. Travel, as a result, is becoming less necessary, and this trend will soon follow into our domestic lives as well.

- What *is* relevant is that we are confronted like never before by cultures and beliefs that are completely alien to us, without further intentional and costly study of a variety of the same.

- Beginning with the foundations for belief is crucial and absolutely necessary to ensure we are dealing with persons of sound judgment.

- The source of truth and acknowledgment of its objective reality brings us all onto an even playing field.

- Without it—moral relativism reigns supreme. Holders of this worldview can and will justify any voluntary act they choose to do for any reason whatsoever. They ultimately determine what is right and good *for themselves*. There are no set universal standards for them. *They* set them.

- Coercion is no better than force in the web of opposing belief systems, and neither are acceptable. Honesty, forthrightness, and concise word usage on the subject at hand ensure minimal confusion and/or misinterpretations between parties.

- Remember that two opposing beliefs cannot both be true by nature. Going directly back to first principles becomes necessary whenever this arises as a conflict between persons or groups.

When artificially intelligent machines and enhanced semi-human cyborgs do begin to mix with the human race, who do you think will eventually become the consummate programmers? Where will this intelligence emanate from? How can one improve on a human intelligence founded on direct acknowledgment of the source of *all* intelligence? The source of computer programming *is* human intelligence…and the source of human intelligence is the One who gave us the ability to possess it.

"Shaping who and what we are and determining how we direct our lives—which includes deciding the ends to which technology will be put—is the traditional role of morality, philosophy, and religion, and the purpose for our social and political institutions."

~ James P. Hogan

The best option is to become actively involved in the process and encourage the same of others. Your interest in this book and subject alone renders you the most eligible and more than likely the best candidate for such a cause to help guide culture in the proper directions.

And if not you—then who? Do you know who represents and carries out your best interests? Get involved.

- Exercise your rights. Don't abuse them by demanding that you be granted them when your use of them is destructive to society and the family unit. Freedom of speech and expression does not imply all speech is right or just, even if expedient for a time. Set the example for others.

The future continues to be a slippery slope of questionable ethics and morals. Techno-gadgets get smaller even before nanotech manufacturing has proven to eventually become the norm in the production of electronics. Implanted chips that give us the ability to "think" surrounding environments into existence and control appliances will render virtual telepathy and telekinesis everyday realities. Nanobots a billionth of a meter small will patrol our bloodstreams, destroying any cells intent on hindering our health and/or accelerating the aging process.

Why someone would want more escapes common sense—but then, good sense ceased to be common long ago.

There will always be abusers of applied science for purely self-indulgent purposes as long as we have rights to privacy and religious liberty...just as substance abuse continues to plague societies. No easy cure—no quick fix—will emerge to meet and abolish these evils.

- Education—teaching truth in the correct historical and present contexts—making time to discuss the meaningful issues to reach mutually understood conclusions...

These will lead us to the answers to our past, present, and future questions concerning humanity's progress. As always, success rests in our own initiative. Others may or may not be open to your newly discovered and developing worldview...some will have to be jarred loose from their very foundation before a new and forever stable one can be laid...

"Know that, yes, some things do happen to you that change you completely.. You can either become bitter or better. You can either live in the mess, or you can turn it into a message that helps other people. You can take disappointments suffered and turn them into divine appointments to help others and minister to their needs."

~ Don Piper

As for those who wonder about alien *War of the Worlds* type invasions...consider such invasion as having already taken place—not from without, but rather from *within*. The aliens are just destroying you from the inside out—*not* from the outside in. If today's cancers are the aliens, then nanobots will be our frontline soldiers—whatever the case, man *is* merging with machine.

I suppose you have a better idea? Well then—we're *all* waiting...

...and so is the world.

Will <u>you</u> rise to the occasion for your generation? Offer more than mere disdain...offer up a solution.

appendixes

GLOSSARY

assuage make less intense or severe; satisfy or appease; pacify

atom smallest unit of an element that can exist alone or in combination

belief mental acceptance of and conviction in the truth, actuality, or validity of something

cliché a trite or overused expression or idea

cyborg term referring to a human being integrated with computer technologies, mechanical parts, and/or synthetic organs grown from stem cells or other means. Essentially, man's attempt to improve on him/herself—including the possible extension of the average lifespan considerably if not indefinitely. [cyb(ernetic) + org(anism)]

cyborgenics study of cybernetic organisms including the social and ethical implications of their existence

demagogue   a leader who obtains power by means of impassioned appeals to emotions and prejudices

democracy   in purest form, the rule of society, governments, and/or implementing laws by majority vote; "mob rule"

dumb   conspicuously unintelligent

egocentric   confined in attitude or interest to one's own needs or affairs; perception from one's own mind as the center

egocentric predicament   the dilemma we face as individual beings that while knowing we are not truly the center of the universe, all things do, in fact, appear to be (and essentially are) occurring all around us at all times

element   substance composed of atoms with the same number of protons

ethics   rules or standards governing the conduct of a person or members of a profession

ethnocentrism   belief in the superiority of one's own ethnic group; an overriding concern with race and leading cause of racism

epigenetics   literally "control above genetics" field of science studying how environmental influences/signals are major determining factors within biological systems. In direct opposition to genetic determinism where genes determine behavior

exponential growth   size or amount increased by a fixed multiple over time (graphically represented as an "I" curve)

femtometer   one quadrillionth of a meter

fractal   geometric pattern repeated at ever smaller and/or larger scales to produce irregular shapes and surfaces (Used especially in computer modeling and observed in irregular patterns and structures of nature)

genetic determinism   overemphasis and/or belief that genetic material is the major or controlling influence on biological systems and subsequently of behavior

genius   of extraordinary intellect and talent; exhibiting exceptionally high intelligence

Hegelian dialectic   theoretical process of arriving at the truth or desired reality by consideration of a thesis, development of an antithesis in reaction to this, and combination resulting in a coherent synthesis (Idealist philosophy of G.W.F. Hegel: 1770 - 1831)

ideology   body of ideas reflecting the social needs and aspirations of an individual, group, class, or culture

idiot   of profound mental retardation having a mental age below three years

idolatry   blind or excessive devotion to something

ignorant   unaware or uninformed; arising from lack of education or knowledge

intellect   ability to learn and reason; capacity for knowledge and understanding; ability to think abstractly and profoundly

intelligence   ability to apply knowledge

law of accelerating returns   as order exponentially increases, time exponentially speeds up—ergo, interval of time between salient events grows shorter as time passes

linguistics — the study of languages and their structure

love — an inward openness bringing greater awareness to the needs of another, subsequently causing a response to appropriately meet them

megabyte — @ one million bytes. One byte equals 8 "bits" or one character (a number, letter, symbol, etc.)

misology — hatred of reason, argument, or enlightenment

molecule — smallest particle into which an element can be divided without changing its physical and chemical properties

moral — conforming to the standards of what is right or just in behavior

moron — of mild mental retardation; having a mental age of seven to twelve years; very stupid

nanotechnology — field involving the building of electronic circuits and devices from single atoms and molecules [**nano** = one billionth or @ five carbon atoms in length when measured in meters]

normal — free from emotional disorder; characterized by average intelligence or development

ontology a branch of metaphysics dealing with the nature of being

oxymoron rhetorical figure of speech in which contradictory terms are combined and presented to render in obvious (but sometimes inconspicuous) nonsensical meaning

picometer one trillionth of a meter

postmodernism broadly inclusive term used to describe the cultural period following "modernism" which was characterized by clear breaks in the traditions of 18th century thought and culture, extended into the late 20th century, and was marked by extraordinary human advancement. The point which the postmodern period begins is widely accepted as when cultural genres became mixed, knowledge became more accessible, and truth is challenged as less objective and held as subjective. Basically a reactionary period to modernism birthed around the mid 20th century.

propaganda systematic and/or material dissemination of information by the advocates of a particular cause or doctrine

prophecy   inspired message by divine influence often viewed as a revelation of divine will but can be derived from intense knowledge and insightful understanding of relevant and/or parallel historical events and cycles

prophet   one gifted with profound moral insight and exceptional powers of expression; chief spokesman of a movement or cause

religion   a set of beliefs, values, and/or practices

republic   rule of social and/or governing order where citizens *entitled* to vote choose representatives to carry out the administration of government on the peoples behalf being fully accountable to them to promote and uphold their focal values

salient   standing out conspicuously

semantics   a branch of linguistics concerning meaning

singularity   a point in time or event beyond which the accelerating rate of changing technologies brought about by machines themselves make all attempts at further forecasting of the state of civilization and/or being— impossible and rather

	meaningless in nature—since humanity will have become extinct
stupid	slow to learn or understand; marked by lack of intelligence
technocracy	government or social system controlled by technicians, especially technical experts within their specialized fields
telekinesis	the movement of objects by scientifically inexplicable means
teleology	the study of design or purpose in natural phenomena; purposeful development, as in nature or history, toward a final end
telepathy	communication through means other than the five known senses
transcend	to pass beyond the limits of
truth	conformity to fact or actuality; genuine reality (by nature exclusively objective)
wisdom	understanding of what is true, right, and enduring in nature; true insight; possessing and practicing good judgment; justly applying knowledge

KEY QUOTES REFERENCE

One of the biggest flaws in the common conception of the future is that the future is something that happens to us, not something we create.

~ Michael Anissimov

Nothing can save us that is possible; we who must die demand a miracle.

~ W.H. Auden

The dullard's envy of brilliant men is always assuaged by the suspicion that they will come to a bad end.

~ Max Beerbohn

Sixty-seven percent of Americans polled say it's possible to believe in both God and evolution. [That God can use evolutionary processes to accomplish His will or sovereign plan.]

~ CBSnews.com (2005)
(bracketed statement added)

Humans evolved their cognitive abilities not due to a few accidental mutations, but rather from an enormous number of mutations acquired through exceptionally intense selection favoring more complex cognitive abilities.

~ Bruce Cahn

Is it progress to upset the whole balance of nature?

~ Prince Charles (2005)

Nothing in life is to be feared. It is only to be understood.

~ Marie Curie

Science without religion is lame—religion without science is blind.

~ Albert Einstein

Great spirits have always encountered violent opposition from mediocre minds.

~ Ibid.

Your beliefs become your thoughts—your thoughts become your words—your words become your actions—your actions become your habits—your habits become your values—your values become your destiny.

~ Mahatma Gandhi

Common sense at the turn of the 21st Century is not worth the price of ink in this statement.

~ R.H.

"New Age" spiritual gurus are egotistical attention starved agnostics—the snake oil salesmen of this generation... They know just enough technical jargon to impress you— just enough for you to let go of your wallet.

~ *Ibid.*

This attempt to make robots employed with A.I. more like humans is ludicrous...We are *inconsistent* beings—at times getting things wrong and even acting contradictory to proclaimed initial intents. We would expect and demand more from these 'smart" machines.

~ Ibid.

A prideful and stubborn character will soon fall—and how great a fall to one who rebuilds on it.

~ Ibid.

A community is like a ship; everyone ought to be prepared to take the helm.

~ Ibid.

Parable of the sower lesson: Prepare the soil before sowing the seed; do what is necessary and so make the message of life effectual.

~ Jesus of Nazareth
from Luke 8:5 - 8, 15 (paraphrase)

. . look not for signs to be observed for it is not "here" nor "there" as some say it is...the kingdom of your Creator is from within you.

~ Ibid.
from Luke 17:21 (paraphrase)

The meeting of two personalities is like the contact of two chemical substances: if there is any reaction, both are transformed.

~ Carl Jung

Human life without death would be something other than human; consciousness of mortality gives rise to our deepest longings and greatest accomplishments.

~ Leon Kass

The primary problems we cannot solve are the ones of which we are not yet aware. For the problems that we *do* encounter, the key challenge is to express them precisely in words (and sometimes in questions).

~ Ray Kurzweil

...increasing order usually means increasing complexity, but sometimes a profound insight will increase order while reducing complexity.

~ Ibid

The future...something which everyone reaches at the rate of sixty minutes an hour, whatever he does, whoever he is.

~ C. S. Lewis

First we build the tools, then they build us.

~ Marshall McLuhan

So now, from this mad passion—
        Which made me take art for an idol and a king...
I have learnt the burden of error that it bore me—
        And what misfortune springs from man's desire...
The world's frivolities have robbed me of the time
        That I was given for reflecting upon God.

~ Michelangelo

Wherever your life ends, it is all there. The advantage of living is not measured by length, but by use; some men have lived long, and lived little; attend to it while you are in it. It lies in your will, not in the number of years, for you have lived enough.

~ Montaigne

We have educated ourselves into imbecility.

~ Malcolm Muggeridge

Work expands so as to fill the time available for its completion.

~ C. Northcote Parkinson

The heart has its reasons that reason does not know.

~ Blaise Pascal

Some people's curiousity is merely vanity. They want to know something in order to talk about it.

~ Ibid.

The role of the infinitely small is infinitely large.

~ Louis Pasteur

Know that, yes, some things do happen to you that change you completely...You can either become bitter or better. You can either live in the mess, or you can turn it into a message that helps other people. You can take disappointments suffered and turn them into divine appointments to help others and minister to their needs.

~ Don Piper

Science cannot solve the ultimate mystery of nature because in the last analysis we are part of the mystery we are trying to solve.

~ Max Planck

Like apples of gold in settings of silver...
Is a word spoken in right circumstances.

~ Proverbs 25:11 *(NASB)*

Most of the greatest evils that man has inflicted upon man have come through people *feeling* quite certain about something which was, in fact, false.

~ Bertrand Russell *(emphasis added)*

Nothing is so common but the wish to be remarkable.

~ William Shakespeare

We are not human beings trying to be spiritual. We are spiritual beings trying to be human.

~ Jacquelyn Small

Nanotechnology has given us the tools...to play with the ultimate toy box of nature—atoms and molecules. Everything is made from it...the possibilities to create new things appear limitless.

~ Horst Stormer

Love is an inward openness to the genuine needs of others, directly resulting in a response of generous and appropriate action in attempts to meet them.

~ Edward Thiele
(paraphrased)

Precision in communication is important, more important than ever, in our era of hair-trigger balances, when a false or misunderstood word may create as much disaster as a sudden thoughtless act.

~ James Thurber

A diamond is just a lump of coal that stuck to its job.

~ Leonardo da Vinci

Within thirty years we wil have the technological means to create superhuman intelligence. Shortly after, the human era will have ended.

~ Vernor Vinge

The Roman Empire fell for many reasons, but three seem particularly relevant for our times: (1) declining moral and ethical values and political comity at home, (2) overconfidence and overextension abroad, and (3) fiscal irresponsibility by the certral government. All these are certainly matters of significant concern today.

~ David M. Walker

I think there is a world market for maybe five computers.

~ Thomas Watson
(IBM chairman, 1943)

Peace without joy is contradictory.

~ Ravi Zacharias

The older we get—the more it takes to fill our heart with wonder—and only God is big enough to fill it.

~ Ibid

To change the meaning of a *word*—is to change the meaning of the *world!*

~ Ibid

FUTURE FORECASTS FROM 2006

- Reminiscent of the pilot's heads-up displays—retinal implants, custom lenses, and/or contacts possessing similar if not advanced functions of telescopic video cameras/recorders and windows-type desktop graphical user interface (GUI).

- Seeming telekinetic ability through brain implants, enabling us to transfer and receive data and images to and from others, as well as from any media source.

- Nanobot construction/demolition of immediate living and working environments by thoughts and specific commands. Full immersion virtual reality.

- Blood cell mimicking nanobots roaming our bloodstreams, destroying various pathogens as programmed, keeping us healthy and subsequently reversing the aging process, and cutting disease off by rooting it out right from the beginning stages.

- Damaged organs either rebuilt, regrown, or replaced by synthetic equivalents, or using lab-grown organs using individuals' stem cells.

- Drug addiction, drunk drivers, and chain smokers eliminated by anti-vice vaccinations, making behaviors intolerable to the former offender/addict.

- Home based and portable nanofactories become the norm, producing custom-designed products from your CAD station or blueprints of items ordered online.

- Television is holographic—3-D images produced by advanced nanosized projectors—and fully interactive, responding to body movements and voice commands.

- Bathrooms equipped with diagnostic sensors alert us to urgent health conditions and make prescriptions and/or doctor appointments if required.

- Homes powered independently through exterior paints that capture and store solar power for current and future needs.

General Sources And Recommended Reading By Author

Anderson, Mark K. "Mega Steps Toward the Nanochip." Wired News, April 27, 2001.

Asimov, Isaac. I Robot. New York: Berkley Books, 1986.

Bell, Gordon. "Ultracomputers: A Teraflop Before Its Time." Science 256, April 3, 1992.

Briggs, John. Fractals: the Patterns of Chaos. New York: Simon and Schuster, 1992.

Broderick, Damien. The Spike: How Our Lives Are Being Transformed By Rapidly Advancing Technologies, revised ed. New York: Tor/Forge, 2001.

Chalmers, D.J. The Conscious Mind. New York: Oxford University Press, 1996.

Clark, Arthur C. 3001: The Final Odyssey. New York: Balantine books, 1997.

Denning, Peter J. And Robert M. Metcalfe. Beyond Calculation: The Next fifty Years of Computing. New York: Copernicus, 1997.

Drexler, K. Eric. Engines of Creation. New York: Anchor Books, 1986.

__________. Nanosystems: Molecular Machinery, Manufacturing, and Computation. New York: Wiley Interscience, 1992.

Dumé, Belle. "Microscopy Moves to the Picoscale." PhysicsWeb, June 10, 2004.

Feinberg, John S. And Paul D. Feinberg. Ethics for a Brave New World. Wheaton, IL: Crossway Books, 1993.

Feynman, Richard. "There's Plenty of Room at the Bottom." Miniturization, edited by H.P. Gilbert, New York: Reinhold, 1961.

Feynman, Richard. What Do You Care What Other People Think? New York: Bantam, 1988.

Freitas, Robert A., Jr. And Ralph C. Merkle. Kenematic Self-Replicating Machines. Georgetown, TX: Landis Bioscience, 2004.

Garfinkel, S.L. "Biological Computing." Technology Review, May - June 2000.

Gates, Bill. The Road Ahead. New York: Viking Penguin, 1995.

Gianaro, Catherine. University of Chicago Chronicle 24.7, January 6, 2005.

Gilder, George. The Meaning of Microcosm. Washington D.C.: The Progress and Freedom Foundation, 1997.

__________. Telescosm. New York: American Heritage Custom Publishing, 1996.

Hara, Yoshiko. "Toshiba Develops Matchbox-Sized Fuel Cell For Mobile Phones." <u>EETimes</u>, June 24, 2004.

Haugeland, John. <u>Mind Design II: Philosophy, Psychology, Artificial Intelligence</u>. Cambridge, MA: MIT Press, 1997.

Huxley, Aldous. <u>Brave New World</u>. New York: Harper, 1946.

James, Mike. <u>Pattern Recognition</u>. New York: John Wiley and Sons, 1988.

Johnson, R. Colin. "Purdue Researchers Build Made-to-Order Nanotubes." <u>EETimes</u>, October 24, 2002.

———. "IBM Nanotubes May Enable Molecular Scale Chips." <u>EETimes</u>, April 26, 2001.

Joy, Bill. "Why the Future Doesn't Need Us." <u>Wired</u>, April 2000.

Kaku, Michio. <u>Visions: How Science Will Revolutionize the 21ˢᵗ Century</u>. New York: Doubleday, 1997.

Kelly, Kevin. <u>Out of control: The New Biology of Machines—Social Systems and the Economic World</u>. Reading, MA: Addison-Wesley, 1994.

Knapp, Louise. "Booze to Fuel Gadget Batteries." <u>Wired News</u>, April 2, 2003.

Knight, Will. "Single Atom Memory Device Sores Data." <u>New Scientist.com</u>. September 10, 2002.

Kulander, Sven and Borje Larsson. <u>Out of Sight: From Quarks to Living Cells</u>. Cambridge, MA: Cambridge University Press, 1987.

Kurzweil, Ray. <u>The Singularity Is Near: When Humans Transcend Biology</u>. New York: Viking Penguin, 2005.

___________ and Terry Grossman, M.D. <u>Fantastic Voyage: Live Long Enough to Live Forever</u>. New York: Rodale, 2004.

___________. <u>The Age of Spiritual Machines: When Computers Exceed Human Intelligence</u>. New York: Viking Penguin, 1999.

Lipton, Bruce H. <u>The Biology of Belief</u>. Santa Rosa, CA: Elite Books, 2005.

Mandelbrot, Benoit B. <u>The Fractal Geometry of Nature</u>. New York: W.H. Freeman, 1988.

Moravee, Hans. "Rise of the Robots." <u>Scientific American</u>, December 1999: 124 - 35.

Naisbitt, John. <u>Global Paradox: The Bigger the World Economy, the More Powerful Its Smallest Players</u>. New York: William Morrow, 1994.

Negroponte, Nicholas. <u>Being Digital</u>. New York: Alfred A. Knopf, 1995.

Norretranders, Tor. <u>The User Illusion: Cutting Consciousness Down to Size</u>. New York: Viking, 1988.

Orwell, George. <u>Nineteen Eighty Four</u>. New York: Harcourt Brace Jovanovich, Inc., 1949.

Paul, Gregory s. and Earl D. Cox. Beyond Humanity: <u>Cyber Evolution and Future Minds</u>. Rockland, MA: Charles River media, 1996.

Pinker, Steven. <u>How the Mind Works</u>. New York: W.W. Norton and Company, 1997.

Reynolds, Glenn H. "The Nano Dilemma." <u>Popular Mechanics</u>, Oct. 2006: 40-43.

Sample, Ian. "Chip Reads Mind of Paralyzed Man." <u>The Guardian</u> (UK), March 31, 2005.

Shachtman, Noah. "A Spy Machine of DARPA's Dreams." <u>Wired News</u>, May 20, 2003.

Smalley, Richard E. "Nanofallacies: Of Chemistry, Love, and Nanobots." <u>Scientific American</u> 285.3, September 2001: 76 - 77.

Steinberg, Douglas. "Determining Nature vs. Nurture." <u>Scientific American Mind</u>, Oct/Nov 2006: 12-14.

Vinge, Vernor. "Technological Singularity." <u>Whole Earth Review</u>, Winter 1993. (Kurzweil AI.net/vingesing)

Zorich, Zach. "Carbon Nanotubes Burst Out of the Lab." <u>Discovery</u>, January 2006: p. 29.

Internet And Other Sources Of Interest

Websites

<accelerating.org>

<cyborganic.com>

<discover.com>

<eetimes.com>

<elevator2010.org>

<en.wikipedia.org>

<extropy.org>

<foresight.org>

<konarkatech.com>

<kurzweilAI.net>

<lhs.com>

<llnl.cov/asci>

<longnow.org>

<media.mit.edu>

<physicsweb.org>

<research.ibm.com>

<research.microsoft.com>

<rzim.org>

<sciam.com>

<static.highbeam.com>

<thefutureoflife.com>

<transhumanism.com>

<transhumanist.com>

<wired.com>

<microvision.com>

<molecularassembler.com>

<newscientist.com>

Books

1984
by George Orwell

The Age of Spiritual Machines
by Ray Kurzweil

Brave New World
by Aldous Huxley

Mind Matters
by James P. Hogan

Nano
by John Robert Marlow

Magazines/Newspapers

American Heritage of Technology and Invention

Business Week

Forbes

Fortune

The Guardian (UK)

NewMan

Popular Mechanics

Popular Science

Science News

Scientific American

Scientific American Mind

The Scientist

Smithsonian

The Week

further inquiries:

Concerning requests about available volume book discounts for civic, church, and other religious organizations . . . speech preparation/assistance services . . . and other special engagements . . .

address your e-mails to:

rahanson273@hotmail.com

subject: cyborgenics*

be sure to include return street address for ground mail response to all serious inquiries

notes

notes

notes